Frank Amasa Bates

The Game Birds of North America

A Descriptive Checklist

Frank Amasa Bates

The Game Birds of North America
A Descriptive Checklist

ISBN/EAN: 9783741186950

Manufactured in Europe, USA, Canada, Australia, Japa

Cover: Foto ©Klaus-Uwe Gerhardt /pixelio.de

Manufactured and distributed by brebook publishing software
(www.brebook.com)

Frank Amasa Bates

The Game Birds of North America

THE GAME BIRDS

OF NORTH AMERICA

A DESCRIPTIVE CHECK-LIST

BY

FRANK A. BATES

President "Boston Scientific Society," and formerly
Associate Editor "Ornithologist and Oologist."

𝔍𝔩𝔩𝔲𝔰𝔱𝔯𝔞𝔱𝔢𝔡

BOSTON
BRADLEE WHIDDEN
1896

S. J. PARKHILL & CO., PRINTERS, BOSTON

INTRODUCTORY.

In considering the preparation of this list, the first point was — "What is a game bird?"

What gunner will admit that his favorite bird is not game, although it be tabooed by his neighbor? And here again another difficulty arose, for many a man shoots Meadow Larks and Reed-birds, and would scorn the imputation that he was not a sportsman, or that these birds were not game.

To avoid one horn of the dilemma without getting hooked by the other, I have manufactured this definition: A game bird is one which is suitable for food and which is habitually pursued by man for sport, demanding skill and dexterity for its capture. I take it for granted that every sportsman is a gentleman, and would not slaughter more game than he could find a use for, and that he would not descend to the level of the pot-hunter, who will kill Robins and other insectivorous birds simply because they are fit to eat.

With this explanation, I apologize beforehand to any-one whose corns I have trodden upon, offering as an excuse, my desire to afford a convenient reference list adapted to the sportsman's needs without compelling him to wander among a mass of useless matter. I have also marked with a star (*) those birds which are not generally accepted as game, although it has been very difficult to draw the line.

KEY TO IDENTIFICATION OF GAME BIRDS.

WATER AND MARSH BIRDS . . . . A.

LAND BIRDS . . . . . B.

A. Toes fully webbed (palmate) . a.
Toes with flaps on side (lobate) . b.
Toes not webbed . . . c.

a. Bill spear-shaped . . . Loons (Nos 1–4).
Bill pointed; saw-toothed . . Mergansers (Nos. 5–7).
Bill and legs long and slender . Avocets (No. 68), Stilts (No. 69).
Bill flat . . . . Ducks (Nos. 8–36).

d. Legs covered with scales (scutellate) .
Legs cross-lined into plates (reticulate) e.

Scutellate

Reticulate

c. Wings short and rounded, flight feeble . . . . Rails (Nos. 65–67).
   Bill long, turned upward . . . . Godwits (Nos. 85, 86).
   Bill long, turned downward . . . . Curlews (Nos. 94–98).
   Length, 3 feet or over . . . . Cranes (Nos. 53–55).
   Toes, 4 . . . . No. 99.
   Toes, 3 . . . . f.

f. Shank covered with scales in front . . . . g. Nos. 107, 108.
   Shank cross-lined into plates . . . . i.
   Shank shorter than bill . . . . Plovers (Nos. 100–105).
   Shank longer than bill . . . . h. No. 81.

g. Toes, 3 . . . . Nos. 71–73, and 81.
   Toes, 4 . . . . Sandpipers (Nos. 75–95).

h. Bill twice as long as head . . . . Nos. 87–89, and 94.
   Bill only a little longer than head or shorter . . . . Nos. 90–92.

i. Legs green or yellow; bill slender . . . . No. 70
   Legs bluish; bill stout . . . . No. 93.

B. Bill long and slender . . . . No. 106.
   Bill spear-shaped, yellow . . . . No. 124.
   Bill short and slender, black . . . . j.
   Bill soft . . . . No. 123.
   Bill short and stout . . . . k.

j. Head and neck naked . . . . Grouse (Nos. 114–122).
   Head and neck feathered . . . . Partridge and Quail (Nos. 109–113).

k. Shanks feathered . . . .
   Shanks naked . . . .

WATER BIRDS.

1. * LOON. GREAT NORTHERN DIVER.
Urinator imber.

Predominating color, a deep greenish-black
in old birds, brownish-black in young, marked
with small white spots. Breast and under
parts white, and a patch of white lines on the
throat. The old birds have a band of white

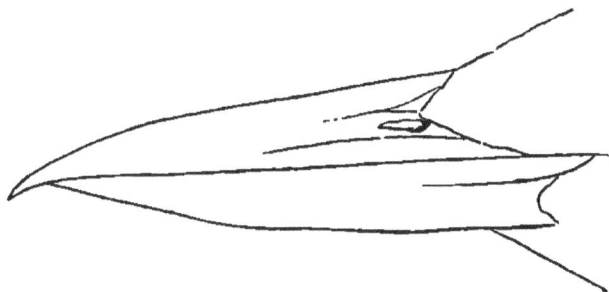

lines about the neck. Iris of old bird, red ;
young, brown. Extent of wings about 4
feet; length from tip of bill to tip of tail,
about 3 feet. The largest one I ever heard
of measured 48 inches in length and weighed
17⅝ lbs. Feet and bill black, the latter
straight and tapering. Legs set far back on
the body.

Habitat — Entire Northern Hemisphere :
in winter goes as far south as the Gulf of

Mexico. Common, in summer, in the large fresh-water ponds; earlier and later in the season in the bays and harbors. Exceedingly wary and difficult of approach, diving at the slightest suspicion of danger and with remarkable powers in this direction.

The flesh has a decidedly fishy taste, but "baked breast of loon" and "loon stew" are esteemed as delicacies by many.

Its congeners are

2. * BLACK-THROATED LOON.

Urinator arcticus.

General color like No. 1, but with less white on throat. Somewhat smaller, measuring 3⅓ feet in extent and 2½ feet in length.

Habitat — More northern than the loon.

3. * PACIFIC LOON.

Urinator pacificus.

Still smaller than No. 2 ; otherwise nearly the same. This bird measures only 2 feet in length.

Habitat — Pacific coast, Alaska to Gulf of California.

4. * RED-THROATED LOON.

Urinator lumme.

About the size of No. 3 and with the same general colors, but the sides of the head and throat are gray, with a patch of rich brownish-red on the throat in the old bird, which is lacking, or nearly so, in the young birds, which, however, are easily recognized by the numerous white spots on the back.

Habitat — About the same as No. 1, but somewhat more confined to the salt water.

5. *AMERICAN MERGANSER.
GOOSANDER. POND SHELDRAKE.
Merganser americanus.

Predominating color, black ; under parts dirty yellowish-white. Head with a slight crest. Wing mostly white. Female of more subdued colors. Iris of eye carmine red in male, yellow in female. Feet generally yellow ; bright red in the male in the spring.

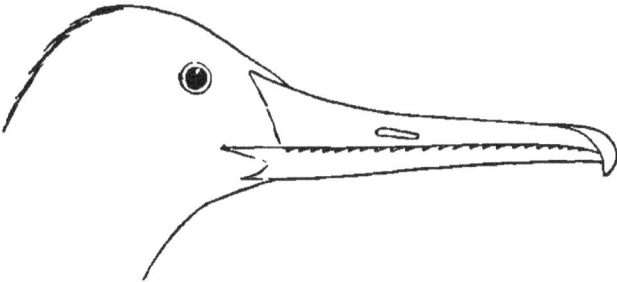

Bill red, saw-toothed on the mouth, with a pronounced hook on the end. Average length about 2 feet; extent nearly 3 feet. The female is much smaller than the male.

Habitat — Entire North America. Breeds
in the northern section; not uncommon anywhere in the ponds and large rivers during
the various seasons, as it migrates south with
the approach of cold weather. They are expert divers, feeding upon fish. Their flesh is
rank, but will fill the belt cavity when nothing else presents itself.

6. * RED-BREASTED MERGANSER. SALT-WATER SHELDRAKE. SAWBILL.

Merganser serrator.

Resembles No. 5, but has a white ring
about the neck of the male, and the crest is
more pronounced. The great feature of distinction is the brownish-red color which
adorns the breast. The female is unlike the
male, but is quite recognizable.
Habitat — Northern sections. Not so common in the United States as No. 5. They
are principally confined to the seacoast, and
their habits are much like the last named.

7. * HOODED MERGANSER.

Lophodytes cucullatus.

Back of male black, under side white; a
beautiful white crest, fronted and bordered
with black, decorates the head. The female
is a little smaller; brownish, with a much
smaller crest, but with much more than No..6.

Eyes yellow. A much smaller bird than the rest of the family, being only 1½ feet long, and a little over 2 feet in extent.

Habitat — North America in general, south to Cuba. This beautiful bird is generally regarded as a prize by gunners, as it is not particularly common anywhere, and the male bird attracts admiration by its graceful carriage and elegant plumage. It is an active diver, feeding upon small fish and aquatic insects, living principally upon the fresh water. Like its relatives, its flesh is but inferior food.

8. MALLARD. GREEN-HEAD. WILD DUCK.

Anas boschas.

Breast of male in spring silver gray, with fine markings of darker. Back black. Head

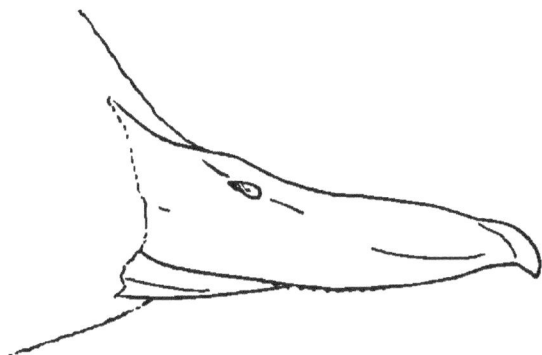

glossy green, with a white band around the neck. The wing bears a beautiful green

patch, framed in black and white. The female is dusky brown and mottled with darker.
In the summer the drake loses his green head,
but can be easily distinguished from the female by the wing patch and the upcurled tail.
Eyes brown ; feet reddish ; bill black, spotted
with orange. Length 2 feet. Extent 3 feet,

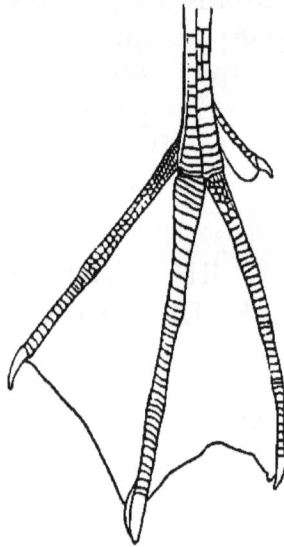

or a little less. Average weight of males 2
lbs. 9 oz., females a little less.

Habitat — The whole of the Northern
Hemisphere. They are not at all common in
New England, but are very plenty on the
inland lakes and rivers, as they are essentially
fresh-water birds.

This species is the progenital stock of our
domestic duck, and has no doubt often been
mistaken for such by gunners. They hybrid-
ize easily with the Black duck, and such

birds make excellent decoys for shooting from blinds. They are easily domesticated, and, wild or not, their flesh is of fine flavor and clean, as they are vegetable feeders.

9. BLACK DUCK. DUSKY DUCK.

Anas obscura.

Predominating color, dark brown; lighter on the breast, head and neck. The edges of the feathers are shaded with dusky, giving the appearance of shell work or scallops. The wing patch is violet; the eyes are brown, the bill yellowish-brown; feet orange, with dark webs. About the same size as No. 8, averaging a little less in weight, and resembling the female of the mallard to a great degree, differing in having no white markings except on the inside of the wings. The male and female are much alike, but the latter is generally of a lighter shade.

Habitat — Eastern North America. This bird favors the salt water more than the mallard, and will be found in immense flocks in our bays and harbors, although it drops into the large ponds during its migratory flights, and is there shot from blinds by means of decoys. Our gunners on the New England coast will insist that there are two varieties, viz, the Red-leg or Southern duck, and the Black-leg or Winter duck, although no lists to-day recognize a difference. My friend Leonard, of Marshfield, Mass., writes me as follows: " In my opinion they are different birds. The Red-leg is larger, tamer, and does not winter here with us. It leaves us here (lat. 42° N.) about the middle of November and does not appear again until the last of February, when they go north. The Black-leg is with us till next May and sometimes breeds here." The Red-leg weighs nearly 3 lbs. The Black-leg about 2¼ lbs.

I shall not attempt to decide the question, for I have no desire to add my name to the list of variety makers (there are too many already), but there is food for thought in the above, coming as it does from one of the best informed market gunners on the coast.

The Black duck is one of the finest table birds, and always commands a call in the markets.

10. FLORIDA DUCK.

MOTTLED DUCK (VARIETY).

Anas fulvigula.

Resembles No. 9 very closely, but the cheeks, chin and throat are more inclined to a buff color, and the light-brown markings of that bird are replaced by the same color. There is a black spot at the base of the bill which the Black duck does not have, and the wing patch is bluish-green; size is about the same. The Texas variety, known as the Mottled duck, is found as far north as Kansas, and differs very slightly.

Habitat — Southern United States. There

are some three varieties of this bird, differing only in minor points, due doubtless to climatic influences. They seem to prefer the fresh water more than do the Black duck, which they so closely resemble that the ordinary sportsman will hardly recognize the difference without close observation, though the distinction is well marked, being somewhat lighter in general color than its near relative.

11. GADWALL. GRAY DUCK.

Anas strepera.

Predominating color, gray with a yellow tinge; back and breast darker, nearly brown. The scaly appearance noted in No. 9 is very noticeable in this variety, as the edges of the feathers are nearly white. Wing patch white. Bill blue-black, in the female lighter and blotched with orange. Legs orange, with dark webs. Eyes red-brown. Belly white, with gray lines. Extent of wings, a little less than 3 feet. Length of bird a little less than 2 feet, being a trifle smaller than either No. 8 or No. 9. Weight about 2 lbs.

Habitat — Northern Hemisphere in general, principally in the interior, as it prefers the fresh water. These birds do not go in large flocks, but they make up for numbers by the noise they make; a small flock will create more disturbance than a raft of Black duck. They are comparatively bold; are clean feeders, and are excellent eating.

12. BALDPATE. AMERICAN WIDGEON.

Anas americana.

Predominating color, gray with fine wavy lines of darker, belly white, as is also the body of the wing, while the tips are brownish-gray. Wing patch green. The body of the head is buff color, shaded with greenish-black and brick-red, and the old drakes have a broad distinct patch of green on the sides of the head, backwards from the eye ; the distinguishing feature, however, is the distinctly white forehead, which is present to a greater or less extent in all the different ages. The females and the young males are darker in general tone and the shell markings are more distinct, while the white crown is less so. There can be no mistake in identifying the species however, from the great amount of white on the wings and under parts. About the same size as No. 11: will average smaller. Bill bluish-gray. Feet a trifle duller. Eyes brown. A very difficult bird to describe, from its great variability in different ages.

Habitat — North America at large, but probably less plentiful in New England than in the Southern States and the Mississippi Valley. In fact the local gunners hardly know the bird.

They are shoal-water birds, and feed upon aquatic plants. They have the curious habit of thieving the succulent deep-water plants from the deep-divers and hastening out of reach before they recover from their surprise. They winter on our southern border, where they congregate in large flocks.

On the Pacific coast we sometimes find a bird almost identical in appearance, but the top of the head is creamy instead of white; there is scarcely a shade of green on the sides of the head, which is cinnamon-red instead of grayish. This is the European widgeon, and is only an occasional visitor. It is reported to be not uncommon on the Pamlico and Albemarle Sounds.

13. GREEN-WINGED TEAL.

Anas carolinensis.

Prevailing color gray: under parts white. Old drakes have a rich chestnut-brown head and neck, with green marking, as in the Baldpate. Breast dotted with dark; wing spot green. The females and young males have dark and light brown as their predominating colors, and sufficiently resemble the old males that there need be no mistake in their identification. Smaller than any of the other ducks, the length being only a little over a foot and the extent less than 2 feet. Bill black; feet light blue; eyes brown.

Habitat — Entire North American continent.

Although small, it is one of our finest game birds and one of the most prolific. They are

found generally in small flocks on the edges
of shallow ponds. They are finely flavored
and a good game bird. The European variety, closely resembling the American, is a
rare visitor to our North Atlantic coast.

14.   BLUE-WINGED TEAL.

Anas discors.

Predominating color brown, under parts
lighter than the upper; head lead-color,
nearly black on crown with a white crescent
in front of each eye. The feathers of the
back have edgings of the lighter color, and
the breast is spotted with the darker; wings
sky-blue with green patch; bill grayish-black;
feet yellow; eyes brown. A little larger
than No. 13. Weight 12 ozs.

Habitat — During migrations this bird covers the greater part of the country east of the
Rocky Mountains. Its habits are very much
like the Green-wing, and like them they are
very swift of flight. While feeding they
are very unsuspicious and can be easily approached with a due amount of caution.

15.   CINNAMON TEAL.

Anas cyanoptera.

The female is very similar to No. 14, but
the male is of a rich cinnamon-brown color;
wings blue as in No. 14, and with green patch.

About the same size as the Blue-wing; bill black; feet orange; eyes orange in male, brown in female and young.

Habitat — Rocky Mountains, north to Oregon. A South American variety, rarely found outside the above limits.

16. SHOVELLER. SHOVEL-BILL.

Spatula clypeata.

Broad-bill; although this latter name is more often applied to the Ruddy duck. Back of male black, with each feather tipped with a lighter color. Head dark glossy green; breast white; belly chestnut-brown with a

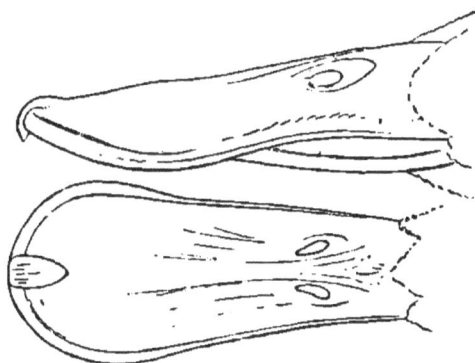

purple shade; shoulders blue; wing spot green, bordered with white; tips blackish; a white spot at each side of the tail; bill brown, and broad at the tip, which is twice as large as at the base; feet and legs red; eyes reddish-yellow. The female and young males are of a soft brown color, shaded with lighter, as is usual in nearly all the ducks. Eyes yellow. Size about that of the Widgeon (No. 12), but a trifle smaller.

Habitat — Entire continent; but rare on the Atlantic coast. Very abundant in Florida in the winter. They delight in shallow pools, where they float, with their heads under water the most of the time, feeding upon the aquatic grasses and insects. Audubon pronounced them as the finest duck for the table in America, even better than the noted Canvas-back. Their flesh is very finely flavored, and, after all, much of the fame of the latter is due to a customary idea.

17. PINTAIL. GRAY DUCK.

Dafila acuta.

Back of male bird gray, marked with wavy white lines; under parts white. Wings dark, speculum green, with coppery reflections. Head and upper part of neck dark brown. The tail is from 5 to 9 inches long, forming a distinctive feature. Bill black, edged with gray; feet gray-blue; eyes brown; size a little smaller than the Mallard, but more gracefully formed. Weight about 2 lbs. The female and young drakes are somewhat smaller, and lack the long tail. The back is brown, mottled with cream-color, while the under parts are light yellow. This plumage is liable to be somewhat varied, as they often cross with the Mallard.

Habitat — West of the Alleghenys; not at all common east, as they prefer the shallow ponds, where the bottoms are covered with herbage. C. J. Maynard says that there are more of these ducks in Florida in the winter than of all other species put together. They seldom or never dive, but stand nearly on end, feet in the air, while pulling the roots from the bottom. They are very swift of flight and quite wary.

18. WOOD DUCK. SUMMER DUCK.

Aix sponsa.

Back of drake very dark brown, belly white, sides marked with gray; breast is reddish-brown with triangular white spots. The top of head and its crest, which is quite long, is dark green, with purple reflections. The throat is white and a section of it branches up toward the eye. A line of white extends backward on each side of the head from the face of the bill and another from just behind the eye. Bill black, with pinkish-white sides, and red at base; feet orange; eyes red. In June the male changes its plumage, and then resembles the female, but by October they have nearly regained their full plumage; during this period, it is said that the drakes flock together. The female would hardly be recognizable as the mate of its gaudily dressed partner. The back is slaty-brown and the belly white; the breast and sides of body are light brown, mottled with dark gray and brown and there is no crest, but the feathers of the nape are elongated.

Habitat — Temperate North America.

This bird, formerly so plentiful, breeding as it did near all the wooded streams of the United States, is yearly becoming more scarce. Ten years ago there was scarcely a stream in New England but had its pair, which lived and reared their young near its banks ; but the beautiful plumage of the male bird was a bait for the sportsman, and the tender and savory flesh for the epicure. In the fall, when they cram their crops to the utmost with acorns, they are in the finest flavor and fit for a monarch. Now these places know them not, and they are plentiful only in the more sparsely settled localities.

19.   RED-HEAD.   AMERICAN POCHARD.

Aythya americana.

Back of male mixed black and white in very fine zig-zag lines, giving it a gray appearance ; belly gray, gradually verging into the color of the back ; wings darker ; breast black ; head and neck rich chestnut-brown ; bill bluish, broad and flattened ; feet gray-blue with dark webs ; eyes orange. The female and young are much like the above, but the head is dull brown, lighter near the bill, and the back is brown, with no zig-zag markings ; eyes yellow ; length a little less than 3 feet ; extent of wings a little less than 2 feet.

Habitat — Arctics to the Gulf. This is like the next. a deep-water duck, culling its food from the bottom and there gathering mollusks, fishes, and submarine plants. It is often mistaken for the famous Canvas-back, which it much resembles and for which it is often sold, as the novice can scarcely distinguish the difference, though it is plainly marked. The bill is broader and more flattened, and blue, with a black belt at end, while in the latter it is blackish throughout.

20. CANVAS-BACK.

Aythya vallisneria.

Resembles No. 19 very much, in fact the Red-head is often sold for this bird, but the head is closer feathered and the bill black, not so broad, and higher at the base : the back is lighter in general color; eyes of male red ; of female red-brown ; the size is a trifle larger than the Red-head.

Habitat — Same range as the Red-head, and its habits apply as well in one case as the other. They were formerly abundant on the Atlantic coast, and while feeding in the Chesapeake Bay on the wild celery (*Vallisneria spiralis*), they attained a flavor which gave them a world-wide reputation as a table bird, which was, in a measure, undeserved. and for this reason were slaughtered in thousands. In consequence they are now getting scarce in this locality, and the principal supply comes

from the Mississippi Valley. These Western
birds are not possessed of the same delicacy
of those from the Chesapeake, and it is a matter of serious doubt whether it is ever the peer
of the Shoveller, much more the superior.
Audubon pronounces in favor of the latter,
while Wilson inclines to the former, but
Audubon was original, while Wilson, with all
his genius, from which I would not for the
world detract one iota, was more swayed by
other's opinions. These birds are famous for
their diving and swimming qualities, and
when wounded will often escape in this way,
oftentimes seizing a sunken root with a death
grip and never arising to the surface.

21. AMERICAN SCAUP DUCK.
BLUE-BILL. BROAD-BILL. RAFTDUCK.

Aythya marila nearctica.

Head, neck and breast black, in the drake,
as is also the rump, tail, and body of wings;
more inclined to a brownish shade in the female. Wing tips lighter than body, and the
wing patch is white; belly nearly white, and
back black and white in zig-zag lines, much
like the Canvas-back: bill blue: feet black:
eyes yellow. The female has a distinct white
face, formed by a white ring around the upper
part of the bill and just back of it. Length
about 20 inches. Extent of wings generally
over 30 inches.

Habitat—North America in general. During the breeding season they frequent the inland ponds and marshes, but at other times they are not uncommon on all the salt-water bays of our coast. They are good divers and feed upon a general course of aquatic diet.

The resemblance of this bird to the next is so close that they are sometimes hardly distinguishable. Opinions vary as to their qualities as food. Coues says they are good when fat. The gunners eat them, but shore men will even eat gulls with gusto ; for my part, excuse me.

22. LESSER SCAUP DUCK.
LITTLE BLUE-BILL.
Aythya affinis.

This bird is not easily distinguished from No. 21, except that it is somewhat smaller, measuring 15 to 18 inches in length, and under 30 inches in extent.

Habitat — Same as last, which it resembles almost exactly, except for being a little smaller. Like them they form rafts or flocks on the water, and when they rise make the air resound with the sound of their wings, but unlike them they prefer the more brackish water of the river mouths, and the great rivers and ponds of the interior.

23. RING-NECKED DUCK.
RINGED-NECK SCAUP.
Aythya collaris.

Bill darker colored than No. 22, brown ring around the neck of drake. Wing brownish-slate, wing patch gray, feet gray-blue, with darker webs. Back nearly black. Female is more like No. 22, but the white face is not so apparent, and it has a white ring about the eye. This bird may also be distinguished from the others by the bill; Nos. 21 and 22 have a plain blue bill, while No. 23 has black tip and base, and a narrow stripe of white defines the central blue band on either side.

Habitat — Generally about the same as the Little Scaup, which it also resembles in its habits. The same remarks apply to one as well as the other.

24. AMERICAN GOLDEN-EYE.
WHISTLER. GARROT.
Glaucionetta clangula americana.

Head and back black, neck, breast and under parts white, with white markings on the wings. Head richly glossed with green. The distinctive feature of the Whistlers is the white patch on the side of the head at base of bill, but not touching it. In this species the spot is round. Length 1½ feet; extent 2¼ feet. Bill black, feet orange with dark webs. Eyes bright yellow. The female has a brownish head without the white spot, and the general tone is duller, and the bird a little smaller.

Habitat — North America in general, and a common winter duck in the United States on both coasts and often in the interior.

A bird of beautiful plumage in the male, an expert diver, fond of the flats.

Opinions differ as to the edibility of their flesh. Coues says : " Meat bad — rank and fishy," but gunners on the New England coast eat them at times, and in olden times roast Whistler was often on the spit, and that when other ducks were more plentiful than now.

25. BARROW'S GOLDEN-EYE.
ROCKY MOUNTAIN WHISTLER.

Glaucionetta islandica.

Habitat — More northern than the last named, although it breeds in the Rocky Mountains. It comes in the winter down the coast and rivers, but is never so common as the other form. It is possible that it is often confounded with the common Whistler, but can be easily distinguished by the white spot before the eye, which is triangular in this species and oval in the other. There is also a crest on the top of the head and a parti-colored bill, while the gloss of the head is purple in the Barrow's and green in the American.

Both birds get their common name from the whistling sound given off by the wings in flight.

26. BUFFLE-HEAD. BUTTER-BALL. SPIRIT-DUCK. DIPPER.

Charitonetta albeola.

Back and head of male black, the latter with a large white patch extending from the eye back to the edge of the nape, forming a continuous patch. The head is very puffy; hair cut *a la pompadour;* neck and under parts white. The female has no puffed head; back brown, under parts dirty white, with white wing patch and on side of head. Bill of drake dull blue; of female, dusky. Feet of drake flesh-color; of female, blue-gray, with dark webs. Eyes brown. Weight about 18 ozs.; female a little less.

Habitat — North America at large. The male of this bird is a beauty and no mistake, but Heaven help the man who tries to eat one. I can eat almost anything when I am hungry, but I prefer to hunger rather than eat a Dipper. They can dive like a Kanaka, and are only excelled by the little Grebe (the hell-diver of the vernacular), which can elude a rifle ball and dodge a sunbeam. Their habits are like their relatives, the Golden-eyes.

27. OLD-SQUAW. LONG-TAIL DUCK. SOU-SOUTHERLY. COCKAWEE.

Clangula hyemalis.

Predominating color of the drake, white; breast, back and wings black, the latter with a patch of long white feathers; head with a patch of gray-blue about the eyes, shading into black toward the back of neck. Two long black feathers protrude from the centre of the tail. In the spring the white feathers of the side are mingled with reddish ones. Bill black, tipped with orange: very light when fresh. Feet blue; eyes red. Female brown on back, shaded white beneath; no long feathers on tail nor shoulders; bill and feet dusky-greenish; eyes yellow. Weight about 2 lbs. A very hard bird to describe, as there are so many variations to its plumage in its different forms. The summer and winter coats are very different in both sexes, but there is no chance of mistake after one of the birds have once been noted.

Habitat — Northern Hemisphere.

Of all the inveterate old gabblers, this bird carries off the palm; a good-for-nothing tattling set of old gossips. The flesh is rank and fishy, as they are fish feeders, but the males are very pretty in their dresses of black and white, decked out with drab and brown. They are often shot for their beauty, but never, I think, for food.

28. HARLEQUIN DUCK.
LORD AND LADY. PAINTED DUCK.

Histrionicus histrionicus.

Sexes very unlike. Predominating color of the drake dull purple, darker on the upper parts than on the lower, changing to chestnut on the sides; marked with white as follows: a patch in front of eye, curling over it toward the crown, where it changes to brown; a round spot on side of head, just below which is a long patch on side of neck; a collar about the neck, and patches on wings and each side of the root of the tail, besides a white crescent on breast in front of each wing; bill greenish-yellow; feet gray-blue with dark webs; eyes red-brown. Predominating color of female brownish-gray, to whitish on under parts; a whitish spot before the eye and behind the ear; bill and feet bluish; eyes brown; length $1\frac{1}{2}$ feet or a little less; extent of wings 2 feet or over.

Habitat — Northern hemisphere, south in winter to latitude 40° north. These beautiful ducks are not common farther south than the coast of Maine on the east coast, and northern California on the west. They are well known to gunners principally from their rarity and their variegated plumage, than from any qualities as a food bird. The female is much smaller, and entirely different in general appearance from the male.

29. LABRADOR DUCK. PIED DUCK.

Camptolaimus labradorius.

Former habitat Atlantic coast, breeding from Labrador northward, descending in winter southward to the Chesapeake. The adult male, a rather large duck, is in general, black, head and upper neck white, with a strip of black on the crown and around the neck; wings pied with black and white. Length about 20 inches. Extent about 30 inches. Weight nearly 2 lbs. Female is a sort of dappled dark gray, lighter on the wings.

This bird is now nearly extinct if not quite so, but only forty years ago they were on sale in our markets, and there is a bare chance that another may yet be taken. Strange as it may seem, this bird has disappeared during that space of time, and now bids fair to follow in the footsteps of the Great Auk. The main opportunity of securing one more of these birds (there are only 38 in existence, and these are valued at about $500 each), seems to be either in finding one among the treasured trophies of some coast gunner of the north, or that one may yet be taken among the flocks which come down our coast in the winter, and this note is written in the hopes that gunners will heed the universal cry and look out for them.

30. AMERICAN EIDER. SEA DUCK.

Somateria dresseri.

Predominating color of drake white; rump, tail and under parts black; top of head blue-black; back of head sea-green. The bill, which is dirty-yellow, runs on each side far up toward the eye. Feet greenish; eyes brown. Length 2 feet and over; extent 3 to 3½ feet. Female very unlike the male. Predominating color brown or tan, barred all over with black.

Habitat — North Atlantic coast; south in winter to Long Island Sound. They principally frequent the rocky shores of our coast, but occasionally are seen on the large lakes. Their flesh is not particularly delicious at all times, but is supposed to be palatable under favorable conditions. The down from the breast is of well-known superior quality.

31. PACIFIC EIDER.

Somateria v-nigra.

Habitat — North Pacific coast, south to California. The habits are much the same as those of No. 30, and the same remarks will apply to both. The distinctive feature which divides them is that this latter bird has a black V-shaped mark on the throat.

32. KING EIDER.

Somateria spectabilis.

Habitat — Circumpolar ; south in winter, in very small numbers, to lat. 40° N.

This bird varies much in appearance from the others of the family, and may be distinguished by the presence of the black V throat mark, from the Eastern variety, and from the Western by the different configuration of the bill, especially in the summer.

33. AMERICAN SCOTER. SEA COOT. BUTTER-BILL.

Oidemia americana.

Color of drake entirely black : bill very peculiar, having a very pronounced hump at the base, which is yellow, changing to black on the tip. Weight 2½ to 3 lbs. The female is smaller, dusky brown, paler beneath, without hump on the bill ; feet dark ; eyes brown.

Habitat — North American coast and Great Lakes.

I have now touched on ground " where angels might fear to tread," for none are so jealous of their favorite bird as the " coot shooter," and there is no family so diversely named and regarded by mankind. A young bird may be a luscious dish for an epicure, and " coot stew " is famous, but an old bird

is simply infamous in flavor, and I never
saw a bird so young as to equal a stew of old
boots flavored with fish oil. Pardon me,
friends, devotees of the wily coot, my educa-
tion has been sadly neglected. I can eat
sculpin, but do not ask me to eat coot.

The females and young of this bird, and
also No. 35 are known to gunners collectively
as gray coot, and weigh from 2 to 3 lbs.

34. WHITE-WINGED SCOTER.
BULL COOT.

Oidemia deglandi.

Plumage of drake black, with white spot
on wing and another under the eye; bill
black, with knob at base, but not so large as
No. 33; eyes white; feet red, with black

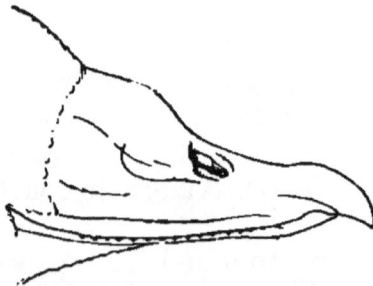

webs.   Female sooty-brown, with same markings, except bill is less bulging, and there is
more white on the head; eyes brown.   Length
1⅔ feet; extent over 3 feet.   Weight 3 to 4
lbs.

Habitat — About the same as the other coots, and resembling them in general habits.

35. SURF SCOTER. SKUNK-HEAD.

Oidemia perspicillata.

Male bird black, with white spot on forehead, and another on back of neck : bill prominent, orange at top and tip, mottled with black at base, a large black spot at base on each side, in front of which is a bluish-white patch ; eyes white : feet orange, webs dark. Female brown, whitish on sides of head and beneath ; bill dark gray ; feet dirty-yellow, webs black ; eyes brown. Weight about 2¼ to 3¼ lbs. A little smaller than No. 34.

This seems to be the most plentiful of this family, at least about the shores of New England, where they are most sought for by gunners, but the comparison is rendered variable by local conditions.

In Massachusetts Bay — the Mecca of " cooters " — they strike in about August 15 and the flight is over about September 20, but the birds stay about all winter. They feed upon the flats, but are then extremely difficult of approach by floating. They are mostly shot outside the harbors as they fly by and are called down by decoys. Curiously enough, although they reach the bay so early, they are seldom seen below Cape Cod until October 1.

36. RUDDY DUCK. BLUE-BILL.
BROAD-BILL.

Erismatura rubida.

Butter-ball. But this latter name must not confound it with the Buffle-head.

Predominating color of full plumaged male, rich brownish-red. Top of head black, throat and sides of head white, belly silvery or gray-white. Generally this bird is brownish, light beneath. Bill and feet bluish. Eyes red-brown. Easily distinguished from ducks of its size, by its broad, flat body, shovel-bill, and short neck and legs. Length 1⅓ feet. Extent less than 2 feet.

It habits the entire continent, and is now better known on inland waters than on the eastern coast.

They are easy of approach but gently slide beneath the water on slight provocation. Expert divers, they swim for some distance beneath the water, reappearing out of gun-shot. They feed upon seeds, roots and shell-fish, and are very edible, being fairly well flavored, fat and juicy. When food is plentiful, they gorge themselves to the utmost, and then are in prime condition for cooking.

37. LESSER SNOW GOOSE.
Chen hyperborea.

38. GREATER SNOW GOOSE.
WHITE BRANT. WAVEY.
Chen hyperborea nivalis.

Color white, wing tips black, head stained with rusty. Bill red, with black on edge of mouth. Legs same color. Eyes dark brown. The young birds look like No. 39. Weight 5 to 6½ lbs.

Habits the western sections and interior of North America, a few visiting the Atlantic coast in the winter, common in the interior.

These two birds are so near alike that I do not attempt to separate any notes on the two varieties. The lesser variety is only about 2 lbs. lighter than the other, and grading up to it, so that it is nearly impossible to distinguish them. Their coloration is identical. The bird which is found in the East is generally, perhaps invariably, the larger form.

In the East they are much in habit like other geese, feeding in the bays and harbors, though I have never heard of their being shot in the ponds. I see no reason why they should not be, as in the West they swarm the prairies to such an extent as to destroy whole fields of wheat.

The flesh is dark colored, and not so finely flavored as some of its relations.

A near form is the

39. BLUE GOOSE.

Chen cærulescens.

Which was long considered but the young of
No. 38, or a colored phase of the same bird.
It is about the size of No. 37, and with about
the same habits. Head white, body dusky
gray-blue, shading into lighter below. Tips
of wings black. Not common.

40. ROSS'S SNOW GOOSE.
HORNED WAVEY.

Chen rossii.

This little goose, not larger than a Mallard,
is an Arctic form, descending in winter to the
lower latitudes. Its habits and characteristics
are but little different from others of the
family.

41. AMERICAN WHITE-FRONTED
GOOSE. LAUGHING GOOSE.

Anser albifrons gambeli.

Predominating color grayish-brown, under
parts whitish, blotched with black; this white
extends backward around the base of tail,
which is black, tipped with white. The dark
color is mottled with brown on the tips of the
feathers. The characteristic mark of the bird
is the white forehead. Bill smooth, pink; feet
yellow, eyes brown. About the same size as

No. 38. The young birds are darker and lack the white forehead, and the bill and feet are darker.

This bird is quite common on the Pacific coast and the Mississippi valley; seldom coming east, although they are occasional visitors. They are more shy than the other geese and hug to inaccessible places very closely. They feed mostly upon aquatic plants and their flesh is of fine flavor.

42. CANADA GOOSE.

Branta canadensis.

Upper parts brownish-gray, shading to a lighter below; bill, head, neck and legs black. A broad white patch on the chin, extending up to nape, and white over and under the tail, eyes brown; extent, 5 feet, length, 3 feet, weight 8 to 15 lbs. when in good condition.

The common wild goose of the entire country, breeding in the north and going south in winter to the Gulf of Mexico. Their migratory flights are strong and rapid, and their V-shaped skeins passing over have drawn out many a farmer to try a shot, and many a one has fallen in this way, and many more have been decoyed to blinds on the shores of the ponds where they had stopped to rest and feed, and never gone on with the rest of the flight. Their flesh is justly esteemed for the table, and their feathers for beds and pillows.

43. HUTCHINS'S GOOSE.
Branta canadensis hutchinsii.

This is a small variety of No. 42, about the size of No. 38, and frequents the western country more than the east. Another variety is the

44. WHITE-CHEEKED GOOSE.
Branta canadensis occidentalis.

Which is a Pacific coast bird; and still another,

45. CACKLING GOOSE.
Branta canadensis minima.

Smaller than any of the others and from the same section.

There is no doubt that all these varieties are entitled to separate mention, but for our purpose they are about the same thing. Every gunner likes to know, however, what his game is, and for purposes of distinction I give Coues's method of separating the varieties: —

Large, no collar of white on neck,		No. 42.
Small, " " " "		No. 43.
Large, with " " "		No. 44.
Small " " " "		No. 45.

46. BRANT. BRANT GOOSE.

Branta bernicla.

Head, neck, back and wings black, some of the quills of latter, whitish on the inside. This color is in some places shaded with brown. Under parts gray, barred with blackish; back of legs it is white, which color extends up over the base of the tail. Sides of neck marked with several white streaks. Bill and feet black, eyes brown. Length 2 feet, extent 4 feet.

This bird habits the Atlantic coast, coming south to the United States, only in the migrations, as far as Florida. They collect in large flocks in the shoal waters, where they feed upon the shell-fish, plants, etc. They seldom dive when feeding, but standing nearly on end, they pull their food from the bottom. They are wary and avoid the shore except when feeding. The sand bars are often black with them while they are dusting. Their flesh is very edible, and they are much esteemed for both food and sport.

47. BLACK BRANT.

Branta nigricans.

Much like No. 46, but the black of the neck runs down on the breast, and the white marks on the neck nearly form a collar. Size of No. 46, and is found on both coasts, but most plentiful on the Pacific; in fact, it is really rare on the Atlantic.

48. *EMPEROR GOOSE.
PAINTED GOOSE.

Philacte canagica.

Predominating color light blue, with wavy
marks of lilac, and clearly defined shell-shaped
markings of black. Head, back of neck and
tail white; throat black, speckled with white;
feet flesh-color; eyes brown.

Habitat—Extreme northwest, coming south
to Alaska, and lower in the winter. A little
larger bird than the Brant. A sea-goose, and
not worth the powder used to kill it for its
edible qualities, as it is rank and fishy.

49. BLACK-BELLIED TREE-DUCK.

Dendrocygna autumnalis.

Predominating color black; head and neck
chocolate, lighter on the chin, white under
tail, on the flanks and on wing; bill red; feet
pink; length 1,$\frac{8}{9}$ feet; extent 3 feet or more.
Lives on the Rio Grande. A very good
market bird.

There is another of these peculiar ducks,
the

50. FULVOUS TREE-DUCK.

Dendrocygna fulva.

Which is yellowish-brown, darker on the head,
no white on wing; black bill and bluish feet;
about the same size and same locality, but
comes farther up the Gulf coast.

51. WHISTLING SWAN.

Olor columbianus.

Color white in full plumage, sometimes with
rusty markings about the head; bill and feet
black, the former with a small yellow spot in
front of the eye; length under 5 feet; extent
6 to 7 feet. Young birds are smaller, ashy-
gray, with a tint of reddish on the head; bill
and feet flesh-color.

They habit the entire continent of North
America, but are rare on the North Atlantic
coast. They feed on aquatic plants and insects, which they glean from the bottom with
their long necks ever and anon poked up in
the air and then thrust down to the mud, but
they never dive while feeding. It is a much
debated question among sportsmen naturalists
whether they ever dive or not; they are, however, credited with this feat when in danger,
but they are seldom approached very closely
without taking wing. Their voice is loud,
and they are exceedingly noisy when feeding,
but it is needless to say that "the sweet war-
bling of the dying swan" is "all in your eye."
They could not warble if they wanted to.

52. TRUMPETER SWAN.

Olor buccinator.

This bird is hardly to be distinguished from
No. 51 except by its larger size, being about
5 feet long and 8 feet in extent of wings,

weighing about 40 lbs. There is one point
of difference in full-grown birds, which should
be constant, viz., No. 51 has twenty feathers
in the tail and No. 52 has twenty-four. Both
these birds are good eating, and sportsmen
need not quibble over the feathers in the tail
if they have an opportunity to bring one to
bag, for they are the cream of bird-shooting.

THE WADERS.

53. * WHOOPING CRANE.

Grus americana.

Old birds white, with black wing-tips: bill greenish: legs black: head naked, carmine color, sparsely covered with hair-like black feathers; eyes yellow. The young birds are grayish-white, with patches of cinnamon-brown; head feathered, of the same brown color. Length 4⅓ feet; extent, 7½ feet, weight 12 to 30 lbs. Female a little smaller. It principally habits the Southern States and the Mississippi Valley. Its voice is loud and can be heard a long distance. These birds feed upon grains, vegetables, worms, mice, reptiles, or in fact almost anything which comes within their reach. They are strong fliers and keep up a continuous croak as they carry their Indian file through the air. They are exceedingly wary, and are best shot from blinds, as they come to feed in the shores of the shallow ponds of the South.

I scarcely am able to distinctively class this bird and its succeeding allies as game birds, although their flesh is not bad eating, and is much relished by some, so I have admitted them under the hypothetical list, as almost any sportsman would drop them if he had an opportunity.

54. * LITTLE BROWN CRANE.
Grus canadensis.

Plumage of old bird lead-gray; wing-tips a little darker. Head nearly bare, as in No. 53, crimson ; bill and feet black ; eyes red. Young, with head feathered, and plumage patched with rusty-brown. Length 3½ feet. Extent, 6 feet and over. The only record of the weight of this bird is one shot in Texas, viz., 11⅓ lbs., but it is stated that they weigh as much as 17 lbs.

The habits of this bird are very similar to No. 53, and, like them, they exhibit strange antics in the mating season. In fact they conduct a war dance, in which the females join as well, and the whooping and circling are equal to a country dance.

There is another bird which is closely related — the

55. * SANDHILL CRANE.
COMMON BROWN CRANE.
Grus mexicana.

Plumage and habits exactly like the last, and in fact it was always considered one species until very lately, but the variety makers have decided that they shall be divorced, basing their decree upon the difference in size of birds. The size of this bird is: Extent, 6¾ feet ; length, nearly 4 feet.

56. KING RAIL. RED-BREAST RAIL. FRESH-WATER MARSH-HEN.

Rallus elegans.

Predominating color olive-brown, streaked with very dark brown ; plain brown on top of head and neck ; chestnut-brown on wings ; reddish-brown below ; lighter on the belly. The sides are darker and streaked with white. Length, 1½ feet ; extent, 2 feet. Bill and feet yellow-brown ; eyes red. Is found in the fresh-water marshes of the eastern United States, south of lat. 40° N., where it skulks and hides in the high grass and reeds, and can even take to the water on approach of the dog, flying principally by night, and not starting easily when approached. They feed upon insects, reptiles and seeds of the aquatic plants and grasses. Their flesh is good, well flavored, and this, combined with their size, makes them a favorite game bird.

57. CLAPPER RAIL. SALT-WATER MARSH-HEN.

Rallus longirostris crepitans.

Varieties —California Clapper Rail, Louisiana Clapper Rail.

General appearance something like No. 56, but considerably smaller, and a little lighter in general tone. The lower parts are more inclined to a grayish color, as, in fact, is the whole bird. Length 15 inches ; extent about 20 inches.

The two varieties are only such as are produced by difference of climate and local surroundings, and are so nearly alike, that the average sportsman could not distinguish them unless laid side by side. It may be very well for our ultra scientific workers to devote their time to finding varieties, and these varieties may be valid and constant, but it seems to the average sportsman that they would be better employed otherwise.

These birds frequent the salt and brackish water marshes of our shores on both coasts, on about the same range as No. 56. They take to the water more freely even than the fresh-water bird, but afford excellent sport if properly hunted. The most approved plan is to place a good poler in the stern of a skiff, while the gunner stands in the bow, and is propelled through the reeds which the birds occupy as a home. This sport requires a quick eye and a steady hand and balance, while the poler must be steady and quick to " mark " the birds as they fall.

In the autumn they are very good eating, as they are clean feeders themselves at this season.

58. VIRGINIA RAIL.

Rallus virginianus.

Not often known to sportsmen as being more than a smaller bird of No. 56, or a dark colored bird of No. 59. In plumage it is almost an exact copy on a small scale of No. 56, measuring about 9 inches in length and 13½ in extent.

They are more or less abundant in the marshes, both salt and fresh, from Massachusetts southward, of course leaving its more northern range when cold weather approaches. They also are found in small numbers on the western coast, and are not rare in the interior. Their home appears to be in the marshes from New York to Carolina, and here they are found in considerable numbers, affording good sport, but there is not meat enough upon them to make them of much value as food. The old saying " as thin as a rail," might have been a comparison with this bird, without deviation from the sense. Our birds in the north in summer are not near so numerous as the succeeding species, but only a few gunners make any distinction, so that perhaps more are seen than are reported.

59. SORA. CAROLINA RAIL. CAROLINA CRAKE.

Porzana carolina.

Predominating color rich yellowish-brown, with numerous streaks and spots of white: head shaded with black, sides lighter, barred with white, and belly nearly white. In younger birds the colors are not so clearly defined, the impression being that of a faded bird. Length about 9 inches: extent of wings about 1 foot or more; bill and legs yellowish-green : eyes brown.

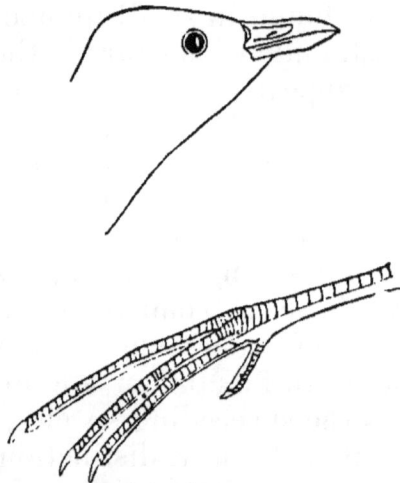

Inhabits the whole of North America; most common in the temperate regions, where it swarms in the reedy marshes. This is the common rail of gunners, and is deservedly a favorite with them. Many are the hours

spent in wading the meadows in pursuit of
these birds. They do not rise much more
readily than others of the family, and have
a way of skulking which is conducive of
bad language.

60. YELLOW RAIL.
Porzana noveboracensis.

Predominating color dull yellowish, barred
with black and white; belly lighter, but with
a more yellowish tinge than any of the other
varieties. Bill darker than the others; feet
and legs flesh-colored: eyes hazel. A smaller
bird, about 6 inches long and a foot or less in
extent. This bird is not common anywhere,
more reserved in his habits, not so noisy, and
moves about mostly in the twilight. It
ranges all over the continent.

61. BLACK RAIL.
Porzana jamaicensis.

Very dark colored, and finely speckled
with white, with some bars; bill black; legs
and feet greenish-yellow; eyes red. Smaller
than any of the others, being under 6 inches
long and about 11 inches in extent. Also
widely distributed, but not commonly seen.

The reason that I note these birds is that
sportsmen may look for them and report their
capture that their relative abundance may be
better known.

The succeeding three birds I admit to this list under protest, as I do not consider them game in any sense of the word, although I would not cast reflection upon any man who shoots them. Perhaps a man may be pardoned for taking that which is pleasing to the artistic sense by reason of its beautiful colors or graceful form, as well as he who caters to the demands of an epicurean appetite. These birds are beautiful : they are sometimes eaten ; they are shot, hence they are here at the solicitation of parties interested.

62.   *PURPLE GALLINULE. BLUE MUD–HEN.

Ionornis martinica.

Color above, greenish shaded with the purple of the head and under parts, darker in belly and wings ; the bill is red, tipped yellow, and above it is a sort of shield of blue color : legs yellow. South Atlantic and Gulf States, sometimes north to lat. 40°. Length 1 foot, extent nearly 2 feet. A beautiful bird with many of the characteristics of the Coots, and like them living in the marshes and on the edges of the ponds.

I found one of these birds, in captivity, in possession of Mr. Andrew Downs, of Halifax, Nova Scotia. It was captured in that place, and when I saw it there in 1889, was reasonably tame. It would feed upon the seeds

thrown in the aviary for the other birds, and
would come down and eat with them. It
passed a portion of the time in the branches
of a small tree, which occupied the centre,
and roosted there at night, placing its long
toes over the spreading twigs where they
forked, as it could not clasp its toes around
the branches.

63. * FLORIDA GALLINULE.
Gallinula galeata.

Back dark slate-color; head, neck and
breast brownish or black, becoming lighter on
the belly; edges of wing white, with stripes
of same color on the sides; bill, which has a
helmet like the last named species, is red with
green tip; legs greenish; eyes red, brown in
young birds. Weight about 1 lb.; a little
larger bird than the last, and inhabiting about
the same country, except that these birds are
regular visitors to the North for breeding, and
they are more inland in their habits. Like
the Rails, they dangle their feet when they
fly, and soon drop.

64. * AMERICAN COOT. MUD-HEN.
MEADOW-HEN.
Fulica americana.

Predominating color slate-blue; much
darker on the head and neck, and tinged with
brownish on the back; edge of wings white,

and same color under the tail: bill white,
shaded with very dark red at tip and base.
Legs greenish slate-color, and the joints of the

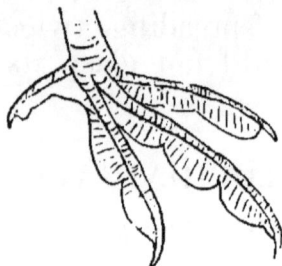

toes are furnished with broad flaps. Eyes red.
Length 1¼ feet, extent 2 feet or over.

Its habits are very much like the Gallinules,
inhabiting the marshes where the reeds are
thick and the water plenty, and though they
delight to sport in the open water, they
quickly take to their reedy coverts when
disturbed.

The reason that I objected to their intro-
duction here was on account of the general
verdict of gunners that they were unfit for
food. I never tried but once; I never shall
try again: a thought of the first experience
is enough, and in my younger days any one
who would eat a "mud-hen" would eat crow.
But I have found many advocates of the bird,
as many as of the "Sea-Coot," another of the
same ilk as regards epicurean tastes, and
hence it is here.

65. RED PHALAROPE. WHALE-BIRD.

Crymophilus fulicarius.

The adult birds, in summer, are nearly uniform wine-color on the under sides, and the top of the head is almost black. The rump is white, and the sides of the head are the same color, which markings extend over the back of the neck.

The back has a tawny appearance, because the black feathers are all edged with brownish. The wing feathers are all marked with white. Eyes brown. Legs black; bill yellowish at base, black at tip. Extent 14½ inches, length 7½ inches.

Sportsmen seldom see this bird in the full plumage, as they are then generally in the far North, and they pass this section of the coast without coming on shore, except by accident, about the first two weeks in May.

In winter they are more common in civilized latitudes, but would hardly be recognized as the same bird. The back is dark gray, and the head and under parts are nearly white. The white markings on the wings are also noticeable.

It is hardly fair to class these birds as shore-birds, as they only come to land when blown in by heavy winds. I think that this bird is typically a sea-bird, and like the Stormy Petrel, makes its bed upon the waves, and sleeps with both eyes open. During heavy northeast storms, I have had it reported

thirty miles inland, but do not remember of
ever hearing from it on other conditions.

66.   NORTHERN PHALAROPE.

Phalaropus lobatus.

Whale-bird.   Adult birds in summer plumage are among our most beautiful birds.   The
back is gray, banded with ochre-yellow.   The
under parts are white, and the neck is nearly
encircled by a band of beautiful brownish-red.
The wings are darker than the back, and
plainly marked with a band of white.   Eyes
dark brown, bill and feet black.   A trifle
smaller than No. 65.   They vary much from
this plumage at different ages and seasons,
the immature and winter birds being darker,
and with the reddish neck less evident.

Much like No. 65 in general habits, living
mostly at sea, where they feed upon the floating weeds which carry the insects and small
crustaceans which they favor.   They occasionally go inland, and I think are found over
the greater part of North America.

67.   WILSON'S PHALAROPE.

Phalaropus tricolor.

Although all the Phalaropes are peculiar,
in that the female is larger and more hand-
somely colored than is the male, in none is it

better exemplified than in this instance. Her neck is of a beautiful purplish-red shading into velvety-black upon the sides of the head, and this color can be traced back to the wings: the back and top of head are gray, becoming lighter at the base of the tail, and the wings are slightly darker. Under parts pure white, breast shaded with buff. Eyes brown, bill and feet black: in size larger than either of the others of the family.

The male is much smaller, and, although he has about the same marks, they are not so bright. The young, and the old birds in winter, are light-gray above, white beneath, and breast shaded darker.

These birds are exceedingly rare in the East, but quite common in the interior and westward.

They are to be found on the ponds and marshes, and feed upon the insects and snails. All these birds are at home upon the water, but seldom or never dive.

This peculiarity of the sexes has given rise to many tales of female supremacy and domination, but I fear that it will follow those of the song of the dying swan, " It's all in your eye."

Although, perhaps, these birds do not come distinctly within the limits of my definition of game, I never saw a sportsman who would not shoot one, and should consider him lacking in mental capacity if he did not.

68. AMERICAN AVOCET.

Recurrirostra americana.

Predominating color of the full plumaged birds is white, shading to a light brown on the neck and head, which seems to fade away, as winter approaches, to a dull gray. Wings black. The bill is extremely long, being about one-fifth the length of the whole bird, curved upward and black. Eyes light brown. Legs very long; dull blue. Extent 2½ feet, length nearly 1½ feet.

This is a very curious bird, with its small body and long legs and bill, and parti-colored plumage. It feeds upon the aquatic insects which inhabit the shallow pools about which it loves to wade, immersing its long bill to the bottom, skimming the top in rapid succession, but they do also sometimes invade the grass fields, and wander over them in their search for food. They are now very rare in the Eastern United States, but seem to be common on the Pacific coast, and the Eastern markets often receive them in consignments of game from the Mississippi Valley.

69. BLACK-NECKED STILT.
LONG SHANKS.

Himantopus mexicanus.

Under side, forehead and base of tail white, and a large spot of same color on each side of the head back of the eye. Top of head, nape

of neck, back and wings black : eyes and legs
red; bill black. The female and young birds
are similar, but not so bright. Length 15
inches (variable on account of the length of
the bill, which is from 2½ to 3 inches long);
extent 2½ feet.

Another long-legged bird, and in habits
and distribution much like No. 68, but more
southerly. I have never seen this bird in the
East, but it has been reported as occurring
here, and doubtless is often in our markets.
They are said to be social, feeding in large
flocks, and not particularly wary.

70. AMERICAN WOODCOCK.

Philohela minor.

There are a number of vernacular names
for this bird, but I think that all will recognize
it without difficulty. It is almost outside the

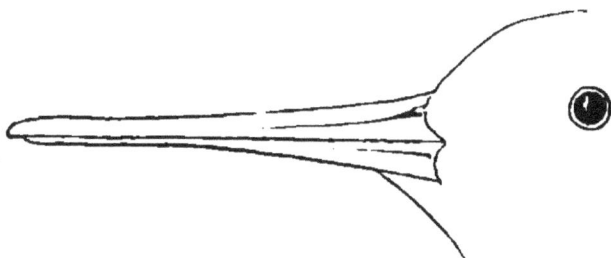

limits of the power of man to describe the
magnificent marking of russet-brown and
black which characterize the plumage of this

bird, which, however, is easily recognizable from the long bill, short legs, stub tail, and the position of the eyes, which are set high up in the head, and far back ; eyes black ; bill and legs flesh-colored. Average weight about 6½ oz., but have been taken weighing 9 oz. The habitat is Eastern North America, north to Nova Scotia, which is now the finest place for this bird within reach of Eastern sportsmen. When I was there in the fall of 1889, big bags were reported during the last of September, and while at Kentville, in the centre of the Province, one man brought in 28 birds taken by himself, and this did not appear to be considered out of the usual course of things. The sportsmen of the Southwest have apparently a bountiful field in the low river bottoms which intersect that country. Many characterize this bird as the " King of Game Birds," and he is entitled to surely a princely rank, if not the head.

There is a bird which has been taken in this country, one-third larger, but with about the same coloration and markings. This is the European Woodcock, and is so very rare that a bird weighing over 9 oz. should be preserved by its fortunate captor, at least until it has been examined by ornithologists. I have a record of a woodcock weighing 12 oz., shot in Maine by a Dr. Gardner, but I have been unable to get more details. If not the European variety, I think it heads the list for weight.

71. WILSON'S SNIPE.

Gallinago delicata.

Known as English Snipe, but this latter name should hardly be used, for the European Snipe is not unknown on this side of the water.

Upper side varied brownish-black and tawny: top of head black, with a tawny stripe down the middle; breast and sides brown, spotted with darker; belly nearly white; eyes brown: feet and bill greenish, the latter about 2¾ inches long; weight 3½ to 4½ oz.; length 11 inches, extent 18 inches. The females will average a trifle less than this, but in all other particulars are similar to the male.

This bird inhabits the fresh marshes all over the continent, and its "Scaipe" and twisting flight is well known. This is probably one of our most difficult birds to shoot, as the moment it is fairly launched into the air it assumes an erratic course of flight productive of misses and profanity. Half-leg deep in the bog, with a crazy bird in front, about the safest plan is to snap at the bird as soon as it starts, and before it fairly gathers itself. This needs a quick eye and finger, and few men are sure of their bird. They are most delicious birds for the table.

72. DOWITCHER.
RED-BREASTED SNIPE. BROWN-BACK.
Macrorhamphus griseus.

Back brownish-black, mottled with light reddish-brown. Under parts of the same light color, shading to almost white on the belly; inside of wings white, marked with dusky; rump white, showing very plainly during flight; tail black, barred with white. In winter the general tone is grayish, mottled with darker; white below, shaded with grayish on the breast, sides and throat; bill and legs greenish-yellow; eyes dark reddish-brown; length about 11 inches, of which 2½ inches is bill; extent 18 inches.

Supposed to inhabit only Eastern North America, and to be replaced in the West by a variety (*M. g. scolopaceus*) which rarely comes East. This latter bird differs only in averaging larger in size; the bill averages longer, and the belly is cinnamon-brown, instead of whitish. The variety makers have decreed that they shall be twain, and they are so laid down; but I doubt if any average sportsman could distinguish them if laid side by side. They are both excellent game birds, and finely flavored, coming to decoys nicely, and alight in a bunch. Those that are not killed when a flock is fired upon, wheel around over their dead and wounded comrades, returning again and again, although repeatedly fired at. They frequent the low flats and marshes, in tide waters following the ebb and flow.

73. STILT SANDPIPER.

Micropalama himantopus.

Predominating color black, mottled with white and brown; a reddish spot is on each side of the head and also a dusky line. Tail gray, under side reddish, mottled and barred with black and white; bill and feet dark green; eyes brown; length 8½ inches; extent about 16 inches; legs very long. In winter they change to a gray color, mottled with lighter, and the legs are paler in color, with the under parts of breast and belly nearly white.

This bird may be considered as really not common anywhere. Their range extends all over the continent east of the Rockies, and the only place where they may be considered as any way common is on the outlying points of our Eastern coast about the first of August, when they roam the flats in company with the sanderlings and tattlers.

74. KNOT. RED-BREAST PLOVER.

Tringa canutus.

Sometimes called Robin Snipe, in common with No. 72. The young are known as gray backs. The adult birds are black on the back, this color being broken by each feather being tipped with dirty white; breast and belly reddish-brown; tail gray, edged with white, short and even: bill stout and greenish-black; legs same color: eyes dark brown; extent 20 inches; length 10½ inches.

One of our handsomest shore birds, inhabiting almost the whole world. In this country most common on the eastern coast, becoming rare westward, about the great rivers and lakes. The young birds are gray, marked with white above, and white, with a tinge of reddish, below. The markings on the back remind one of a succession of black and white semicircles. The spring migration passes lat. 40° N. about May, and returns early in August, at which time they are very fat, and a *bonne bouche* for an epicure. They feed on shell-fish and marine crustaceans, which they pick up on the flats at low tide. They follow the flow of the waves upon the beach, running back and forth vivaciously, and not seeming to mind when they were not quick enough to avoid an incoming wave, which took them from their feet and floated them along. The call of two sharp notes — "Wheep, wheep" — is easily imitated, and they will often come within range, without blinds or decoys.

75. PURPLE SANDPIPER.

Tringa maritima.

Predominating color of the bird which is generally known to sportsmen, is dark purple on upper parts and wings, the latter edged with white : breast lighter, each feather edged with white. This color extends along under the wings, but the belly is pure white ; legs and bill flesh-color, the latter with black tip. The adult breeding plumage is seldom seen, as the bird has a circumpolar range, coming South only in the autumn and winter. Their length is 9 inches, and extent about 16 inches. They are rarely seen before the first of October, and confine themselves to the rocky beaches, so that only a few have ever seen them, and still fewer have shot them.

There are two birds closely allied to this one, but they probably never come farther south than Alaska. I never saw one, and know nothing about them. They are the Aleutian and Prybilof sandpipers.

76. PECTORAL SANDPIPER.
GRASS-BIRD. JACK-SNIPE.

Tringa maculata.

The color of the upper parts is a mixture of ashy and reddish markings on a dark brown ground ; under parts white : breast and sides of neck dull light brown, marked with streaks of darker: patch of white under the chin, and a white line over the eye. The crown is a mixture of streaks of dark brown and light chestnut, and the nape of the neck is lighter, being streaked with two shades of a dull yellowish cast. Wings darker than body. In the young fall birds the breast has more of a yellow tinge. Base of tail black; eyes brown: legs greenish-yellow. Length 9 inches; extent 17 inches.

These birds are found all over the country, and I think are, as regards habits, pretty much the same wherever found. They love the low, muddy shores and flats, and the grassy meadows above tide water, feeding upon the crustaceans, grasshoppers and other small insects.

They are a favorite game bird, although small; as they are quick of flight, somewhat resembling the snipe, and when " walked up" give a good shot an opportunity to show his skill. They are fine food, although small, and one never need throw them one side as useless.

There are two other birds closely allied to this, and to the casual observer differing but little. The first,

77.   WHITE-RUMPED SANDPIPER.

Tringa fuscicollis.

Differs principally in the base of the tail being white; the breast without the dull color; and smaller size. Length 7½ inches, and extent 15 inches. Eyes brown. legs brownish. Found on the beaches as well as on the marshes.

Gunners on the east coast have found a smaller bird mingling with flocks of No. 76, and they often have remarked that they are a different species. I have never been able to secure one of the smaller birds, but have no doubt that it will prove to be

78. BAIRD'S SANDPIPER.

Tringa bairdii.

Rare on the coast but plentiful in the interior, and differing from No. 76 in its smaller size, the length being 7 inches, and extent of wings about 15½ inches. The neck is the same color as the crown. The markings on the back have a tendency to yellowish rather than chestnut, and red as in the Grassbird. The shading on the breast is light, and the bill and feet are black.

79. LEAST SANDPIPER. PEEP.
BUMBLE-BEE.

Tringa minutilla.

Back of full plumaged birds ashy-gray, mottled with black and brown; throat and sides, grayish : under parts white : eyes brown : bill and feet, greenish-slate ; length, about 5½ inches ; extent, 11 inches. This bird hardly

needs especial description, as its small size
is sufficient to distinguish it from the others,
although No. 82 is but little larger.

In habits, this species is but little different from the rest of its tribe, running about
among the pools of water left by the receding
tide, picking up the insects and shell-fish.
They are exceedingly restless, and seldom in
one spot for more than a moment. They breed
in the far North, but are so irregular in their
movements that they are with us nearly every
month in the summer.

I am much in doubt whether to admit these
birds in the list, but they are so closely asso-
ciated with others of " like ilk," that I am in
a measure obliged to describe them. It
would take about a dozen to make a mouthful, but when that mouthful is obtained, it is
one of the sweetest morsels that ever titillated
the palate of an epicure. My mouth waters
at the thought of "peep-stew," and I can
pardon the man who calls them game, though
they be not larger than sparrows.

80.   RED-BACKED SANDPIPER.
DUNLIN. FALL-SNIPE. OX-BIRD.
Tringa alpina pacifica.

In the full summer plumage, the back is a
reddish-brown, mottled with black and shaded
with grayish touches ; wings, mottled gray
and brown, shaded with white ; the head,
neck and breast are ashy, marked with elon-

gated spots of darker; belly black, rest of under part white.

The winter plumage is so much different, that they are often esteemed as separate birds. They then lack the red back and the black belly. The upper parts are mottled gray, and the under parts nearly white, the breast being streaked with dusky: eyes dark brown; bill and feet black; length, 8½ inches: extent, 15 inches.

This is deservedly a favorite bird with sportsmen, both from its beautiful plumage, and for its edible qualities. They are on our New England shores about the first of May, and again during October. They feed on the sandy flats, and in the autumn are easily captured, any boy being able to walk them up or call them down. They inhabit the whole of North America, breeding in the Arctics.

There is a slightly smaller bird, which can be only distinguished by its size, and disproportionately elevated bill. This is the European Dunlin, a rare visitant to our Atlantic shores.

81. CURLEW SANDPIPER.
FERRUGINEOUS SANDPIPER.

Tringa ferruginea.

Top of head and back, bright greenish-black, mottled with a clear reddish: neck and under side, reddish-brown: bill and legs,

greenish-black; the former long and curved
downwards, and the latter long and slender.
This bird is a very rare visitor to our Atlantic
shores, and there are only about fifteen instances of its occurrence here recorded. It is
with the hope that more may be reported
that this note is inserted. The length is
about 8½ inches, and the extent about 14
inches, nearly the size of the Dunlin.

82. SEMI-PALMATED SANDPIPER.
PEEP.

Ereunetes pusillus.

Predominating color grayish-brown, formed
by the intermingling of black in the field of
each feather, surrounded by red and tipped

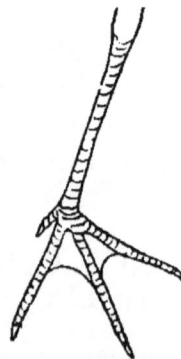

with white; pure white below.   In winter
and in the young birds the color is grayish, and in both dresses the breast is shaded

with dark. Length a trifle over 6 inches, extent about 12 inches; eyes dark brown; bill and legs greenish-black.

This bird much resembles the other peep (No. 79), and probably but few sportsmen recognize a difference between them, and for their purposes no distinction is necessary, as their habits are similar.

Many a sportsman, when the larger birds are not flying, makes out his day's sport with these little birds, and when they are wild it will require some patience and craft to secure more than a dozen or so unless they are very plentiful.

83. WESTERN SANDPIPER.
Ereunetes occidentalis.

This is another case of hair splitting on the part of our ornithologists, and I do not think that over one in fifty of my readers could distinguish this bird from the last if they lay side by side, so for all readers west of the Rockies, for No. 82 read No. 83.

84. SANDERLING. GRAY-BACK.
BULL PEEP.
Calidris arenaria.

In the summer plumage the back is mottled with shell-like markings of black, gray and reddish, formed by each feather having a black centre and a reddish or gray tip. This

marking is a very difficult one to describe in
unscientific language, and must be seen to be
appreciated. The belly is pure white and
the breast mottled. The winter and young
plumage lacks the reddish and the breast is
shaded with buff. Length 8 inches; extent
over 15 inches; eyes brown; bill and feet
black.

This bird is found at various seasons over
the entire continent, and for beauty is sur-
passed by but few. It passes Massachusetts,
going northward, early in May, and returns
in July, the adults coming first and the young
following in August. They feed upon the
sandy beaches and flats almost exclusively,
and seem to be in constant motion. They
will follow a "breaker" down until they are
wading in the rolling surf, and skurry back
to avoid the rush when it returns.

Since the approach of civilization has
driven the larger birds away from our shores,
more of these birds are shot than formerly,
but they are as deserving as the plovers of a
place in the game bag. They are a good
table bird, and are generally in good condition, especially the young birds in the fall.

85. MARBLED GODWIT. MARLIN.

Limosa fedoa.

Predominating color dark brown, variegated with reddish-brown spots and markings. The under parts are of a rusty color, and the markings are finer. The bill is long, curved

upwards slightly, flesh-colored, with dark-brown tip. Legs long, blackish. Length 18 inches; extent about 33 inches. Eyes brown.

Distinguishing Features. — Rump and upper side of tail barred; lining of wings reddish; markings on under side in short streaks; bill 4 to 4½ inches long.

This, one of the largest of our "shore birds," is not very common on our Eastern shores, but will be found in goodly numbers south of Cape Hatteras, where they congregate on the marshes; in the interior they frequent the wet prairies. They are shy birds, but come to decoy well, and, if one is wounded, the others will often hover over, affording another shot, if the gunner is not in too much of a hurry, and will call them back.

There is another bird of this family, which is common in Alaska, known as the White-tailed Godwit, but it so seldom comes within our range that I do not give it especial mention. It is common in the Old World.

86. HUDSONIAN GODWIT. SPOT-RUMP.

Limosa hæmastica.

Predominating color more of a grayish cast than No. 85; and the under parts are reddish-brown, barred with dusky and whitish lines. The winter plumage is gray, and under parts grayish-white, with but few markings. Bill like the last-named bird, but much shorter ($2\frac{3}{4}$ to $3\frac{1}{2}$ inches). Legs slate-colored; eyes darkish-brown. Length 15 inches; extent 27 inches. There are variations of plumage in every combination between the two stages.

Distinguishing Features. — Tail black, tipped with white, and with white base, giving the common name of "Spot-Rump"; lining of wings blackish.

This bird is somewhat smaller than the Marbled Godwit, but the habits are very similar. Its range is over the entire continent, south to South America, except west of the Rocky Mountains, but it nowhere seems to be common.

Both these birds are delicate in flavor, and are true game birds. They are often confounded with the curlews, but are readily distinguished, as the bill turns up, and in the other it turns down.

87. GREATER YELLOW-LEGS.
WINTER YELLOW-LEGS. TATTLER
Totanus melanoleucus.

Back nearly black, speckled with white.
Head and neck lighter, due to the white being in the form of streaks; a white line
above the eye, and the rump white, slightly
barred. Under side white, marked with a
few lines toward the breast, which is much
streaked with gray. Eyes brown. Bill black
and long ($2\frac{1}{4}$ inches), slightly turned upwards.
Legs long and slender, chrome-yellow in color.
Length about 13 inches; extent 2 feet.

The young plumage is lighter, with fewer
dark markings.

These birds arrive early from the South,
are gone to the far North before summer opens,
and on their return linger in the temperate
latitudes until the cold weather is fairly upon
us.

I know not whether this bird is the more
admired as a game bird, or hated by gunners
for its noisy, vociferous cries when it sees
them: many a good shot at plover has been
spoiled by one of these birds setting up its
infernal "cu-cu-cu-cu-cu-cu" when they were
crawling upon a flock of birds.

Still they are not difficult to secure; are
easily called to blind on the edge of the sedge,
as their call is not difficult to imitate.

They feed principally upon the small fish
which swim about in the little pools left upon

the flats by the receding tide, and they wade
about in the shallow water, following its edge
as it recedes and advances, continually keep-
ing up a bobbing motion, as if they kept up
the motion so as to strike more quickly when
occasion served.

88. YELLOW-LEGS.
LESSER YELLOW-LEGS.
Totanus flavipes.

A miniature counterpart of No. 87 so far
as all practical purposes go. Length 11 inches;
extent 20 inches; bill 1½ inches and straight;
the legs are longer in proportion to its size
than in No. 87. It is found in the same
localities as its larger prototype, and resembles it in habits. In New England, however,
it is known as Summer Yellow-legs, since it
arrives from the North about the middle of
July, and leaves for the South as early as
September 15. It is, in autumn, a fat, fairly
flavored bird, but is so easily captured that
one soon tires of shooting them. They will
come again and again to the blind, where
lie outstretched a number of their fellows,
who have just fallen, and do not seem to
mind the showers of lead which are poured
into their midst.

I have often wondered where they passed
in their spring migration northward, as we
seldom see them at that time upon the New

England coast, and I am inclined to think that they go up the valley of the Mississippi, and spread out after they pass the Canadian range of mountains.

89. SOLITARY SANDPIPER.

Totanus solitarius.

Above, greenish-brown, finely speckled on back and streaked on the head and neck with white; white on the under side, the sides of the body and neck marked and barred with dusky; tail barred white and black; bill black; feet greenish; eyes brown. The shades are lighter in the younger birds. Length 8½ inches; extent 16½ inches.

It seems strange that this bird does not figure more in the more popular treatises on game, for it is not an uncommon visitant to all sections of the country. It prefers, I think, the fresh water, although often found feeding from the pools of salt water near the shore. They are very shy, and more or less like the little "Tip-up" in their behavior, bobbing about on the shores of the shallow streams and ponds. They are very quiet and reserved, and seldom whistle except when started.

90. WILLET. HUMILITY.

Symphemia semipalmata.

Predominating color of old birds in full
plumage dark ashy-gray, varying in shade with
the age, and more or less marked with blackish, in interrupted bars; there is also a shading of brown; under side white, with brownish shade, and marked on the breast and
sides with black; rump white, and there is a
white band on the wings which is very apparent during flight; eyes brown; bill black;
legs bluish.   The young plumage is light
ashy-gray, under side white, and there are
numberless intermediate shades between this
stage and full plumage.   Length 16 inches;
extent 28 inches.

It is claimed that there is a variety,

91. WESTERN WILLET.

Symphemia semipalmata inornata.

Which occurs in the district west of the Mississippi.   The distinguishing trait is only
apparent in full plumaged birds, except
for the slight difference in size, the western
bird being a trifle larger, bill longer and
slenderer, general tone lighter, and markings
not so apparent.

I am not going to quarrel over a shade in
color, but whether you find the bird in Kansas or in New Jersey you will find him the

same shy, suspicious yelper who starts into
the air at the first sniff of danger, sounding
his alarm note, to the end of warning all the
birds within hearing, and discomfiting the gunner. They will, however, come to stool in
very fair shape, but when secured are not of
much value, as they are tough and of not
particularly good flavor. Their haunts and
food are much like No. 87.

92. WANDERING TATLER.

Heteractitis incanus.

Predominating color dark gray, under side
white, sometimes shaded on the throat, and
in other phases of plumage streaked and
barred with dark. Bill black, length 10 inches, extent about 18 inches. This bird is
found along the extreme Pacific coast, and I
know nothing about it, and never saw but
two specimens.

93. BARTRAMIAN SANDPIPER.
UPLAND PLOVER.

Bartramia longicauda.

Predominating color dark brown, each
feather being edged with reddish. The neck
and breast are lighter, which color extends to
and upon the head, the top of which is dark

brown. The breast bears the V-shaped mark
characteristic of many of our shore birds.
Under parts dirty white. The tail is barred
with black, the centre feathers being darker
than the others. Eyes hazel-brown. Bill
yellow below, black above and at tip. Legs
greenish-yellow. Length 12 inches. Extent
22 inches.

These birds differ from the majority of its
kind in frequenting the high fields, pastures,
and prairies, in their search for food, which
consists of grasshoppers, crickets, etc.; and
it is seldom seen on the shores.

They are exceedingly variable in their
spring migrations, arriving here (Lat. 42° N.)
from the middle of April anywhere along until the same time in May, and then perhaps
there may only a very few appear, where the
season before they were plenty. They return
in July, and are with us about a month.

Here in the East a more wary bird does not
exist, not even barring the grouse, and a
good bag of these birds is a rarity.

I have heard that on the Western prairies sportsmen ride them down, and no doubt
many a gunner here wishes he could, as he
sees a scared flock go " over the hills and far
away " before he can get within range.

They must be stalked, or, if the gunners
work in pairs, they may often be driven over.

They are delicious eating, and justly much
esteemed.

94. BUFF-BREASTED SANDPIPER.

Tryngites subruficollis.

Prevailing color mottled dark-brown with a greenish shade, each feather above being tipped with yellowish. The distinguishing feature, however, is the buff-colored breast and belly, unmarked save for a few spots, where it commences to shade into the darker color of the back : this color is deeper in the spring, and in adult birds. Length 8 inches. Extent 16 inches. Eyes brown. Bill dark brown. Legs yellowish-green.

This bird is rare along the Eastern coast, and I think common nowhere. It seems to fly in company with No. 76, and, being about the same size and of similar habit, I have no doubt is often taken for that bird in the fall migrations.

95. SPOTTED SANDPIPER. TEETER. TIP-UP. OX-EYE.

Actitis macularia.

Color olive-gray, finely mottled with black. Under side white, spotted with black, and there is a white mark over the eye. Length 7½ inches. Extent 13 inches. Eyes dark brown. Bill flesh-color, tipped with black. Feet pinkish-white.

I hardly know whether to class this bird as a game bird, but it is hard to say "where

the chicken ends and where the hen begins."
It is as large as a Peep, Nos. 79 and 82, and
many a sportsman has made out an otherwise
meagre bag with these birds, so why not
" Teeters "? They come here in April, and
stay all summer, feeding and breeding on the
shores of the ocean, and, in fact, almost any
small pond or river which will afford them
food.

The motion of bobbing the tail while
moving or standing is as characteristic as is
the bobbing of the head indulged in by the
Tattlers.

I have never seen these birds take to the
water, but have been told that they will
even swim under water for a short distance.
They seem to prefer the calm pools to the
surfy shores.

I cannot say how they would do for eating,
but should think they would be very good
if enough could be obtained to make a showing.

96. LONG-BILLED CURLEW.
SICKLE-BILL.

Numenius longirostris.

Predominating color reddish-brown, barred
or spotted with darker. Under side cinnamon-brown, unmarked save for a few faint
streaks of darker. Length about 2 feet. Extent over 3 feet. Eyes dark brown. Bill black,

the under side yellowish, and very long, much curved downward. This is a distinguishing feature of the bird, often measuring 6 or 8 inches in length. Legs bluish. Weight averages about 30 ozs.

This is one of the largest game birds which visit our shores, and is accordingly much prized by sportsmen, although as a table bird

it is far inferior to the Plovers. They will stand a heavy dose of lead without coming down, and as they fly high, are not always captured when seen.

When they were more common, they were easily "whistled" down, but they have become rare on our shores, although not uncommon in the Southern States in winter.

They feed on the flat sand-bars upon crabs and other small crustaceans, which impart to their flesh something of their characteristic flavor.

97. HUDSONIAN CURLEW.
JACK-CURLEW.

Numenius hudsonicus.

Very similar in general appearance to No.
96, but smaller and lighter in color. More
gray in general tone than reddish, and whitish underneath. Length 18 inches. Extent
32 inches. Bill black, and much shorter in
proportion than that of No. 96, seldom being
over 3½ inches long, but with the decided
downward curve peculiar to the family.

This is our common Curlew, although the
Dough-bird, No. 98, is most often seen in the
market, being more prized for its flesh; the
"Jack" being ranker, more like its larger
relative, No. 96.

The flight of any of the Curlews is not
very rapid, but is strong and well sustained.

98. ESKIMO CURLEW. DOUGH-BIRD.

Numenius borealis.

Very similar in appearance to the other
Curlews, but smaller than either of the others,
and of a generally richer tone of color than
No. 97, from which it varies in having V-
shaped marks on the breast instead of short
bars; but the size of the bird is the best
point of difference. Length about 13 inches.
Extent 28 inches. Bill blackish, about 2¼

inches long, and more slender than in the others of the family. Legs dark blue, with a greenish shade. Eyes dark brown.

This bird is found farther from the shore than the other Curlews, as it loves the dry marshes, and the fields and pastures along the seashore, where it can find its favorite food of grasshoppers and crickets, and the higher land berries, when these are to be found, and which impart to their flesh the more pleasing flavor than is characteristic of the other Curlews, and which renders them of more value as a table bird, hence commanding a higher price in the market.

There has been much debate over the orthography of its common name, *Dough* or *Doe*-bird. I have given preference to the former, merely from the individual opinion, backed by that of local New England market gunners, based on the description of its fat condition, which conveys the idea of a lump of dough.

99. BLACK-BELLIED PLOVER.
BEETLE-HEAD.
Charadrius squatarola.

In full breeding plumage, the upper part of the head and neck is nearly white, with faint gray markings; the back is nearly black, mottled with white and brown. Tail barred with black and white. The sides of the head, throat, and under side of body, is deep black,

shading into white at the base of the tail.
Feet and bill black. In the immature and
winter plumage all this is changed, however,
and the bird loses nearly all its black color;
the back becoming mottled brown and white,
the breast nearly white, with a few darker
markings; even the legs and bill assuming
a lighter tone; and these birds shot at the
different seasons of the year assume all the
various grades between these two plumages.

These birds are often confounded, more
particularly in the gray phase, with No. 100,
but they are easily distinguished by the presence of the minute hind toe, which is lacking
in the Golden Plover (see No. 100).

Extent of wings 25 inches. Length of bird
11½ inches. Eyes brown.

This bird is found in nearly every quarter
of the globe, and migrates North and South
with the changes of the season. The flocks
frequent the beaches at low water, and, when
the flats are covered, they retire to the
marshes, there to await the fall of the tide to

uncover their favorite food of small shell-fish, although they eat the insects of the higher beaches.

They are shot from blinds at high water, over decoys, and whistled down, as they fly from the flats as these are covered by the flowing tide. But the gunner must be a good "caller," or his bag will hang light, for an old "beetle-head" is as crafty as a fox, and an experienced gunner will not even use decoys on the high beaches, as the old birds will often not come down to them, but he will pick out a spot which was occupied, on a former tide, by the birds, dig a hole in the sand, cover it with boards and sand, leaving openings to shoot from and knock them over as they come up.

On the flats a sink-box is built on a bar, either natural or artificial, and the birds will approach, as it is uncovered before the surrounding flats are bare.

I know of no bird which is a more universal favorite with sportsmen, as it requires the greatest skill for its capture, and affording, as it does, a fitting reward for the trouble undertaken.

The flesh is delicate, and the birds generally plump.

100. AMERICAN GOLDEN PLOVER.
GREEN-BACK. BULL-HEAD.

Charadrius dominicus.

Prevailing color brownish-black, in the old birds, marked on the back with numerous round spots of golden yellow, under parts black. This plumage changes in the autumn to a duller shade on the back, and the black disappears from the breast, which is now dirty white, irregularly blotched with dark brown spots which disappear in the winter, when both old and young birds are dark brown above; the light markings are larger than in the spring, giving a generally lighter tone.

The under parts are now dull white, with grayish spots, which wholly disappear towards the base of the tail. Length about 10 inches. Extent about 22 inches, being a little smaller than No. 99. Feet and bill dark bluish. Eyes dark brown.

This bird is often confounded with the Beetle-head (No. 99), but is easily distinguished by the absence of the little hind toe. There are many other points of difference, as may be seen by comparison of these notes, but this is always constant and easily apparent.

It shares with its relative (No. 99) the esteem of sportsmen, and rivals it in its delicacy as a table bird.

It frequents the high beaches and neighboring uplands, and in habits and food much resembles No. 98, with which it much associates.

These birds are shot on their flights, from holes dug in the ground and used as a cover for the gunner.

101. KILLDEER.

Ægialitis vocifera.

Top of head and upper parts dark grayish-brown. Forehead, stripe over the eye, and under parts white, with two black bands on the throat. Rump bright chestnut. Wings and tail variegated with black, brown and white. Length 10 inches. Extent 20 inches. Legs yellowish-gray. The immature plumage is similar but not as bright.

They frequent the pastures and marshes and the borders of muddy ponds, and are very abundant in the West. They were formerly very plentiful on our Eastern shores, but of late are seldom seen. The only time that any number have lately been seen East, was in November of 1888, when thousands were driven upon our shores by the great gale of November 25, and was then announced by me in the *Boston Transcript* of December 12. It was thought by some that these birds were now to return to their old haunts, but

they have never been seen since in any
numbers.

In the West, where they are abundant,
they are easily found, as their vociferous cries
of "kill-deer, kill-deer," are constantly ut-
tered, and they are not difficult to approach.

102. SEMI-PALMATED PLOVER.
RING-NECK.

Ægialitis semipalmata.

Upper parts dark gray. Under parts white.
A broad black band encircles the neck, bordered in front with white, and a white stripe
on forehead surrounded by black. Length 7
inches. Extent 15 inches. Bill yellow, black
tipped. Legs flesh-color. Eyelids orange.
Toes partly webbed.

I presume that I may be criticized for including this among game birds, but it affords
good sport when larger birds are scarce, and
is good eating.

They frequent the flats and beaches, going
to the higher beaches to roost.

103. PIPING PLOVER.

Ægialitis meloda.

Pale and ashy-gray above. Forehead, sides
of head, under side and ring around neck
white. There is a black bar across the top of
head between the eyes, and another which

more or less encircles the neck. (There is a variety in the West, occasionally found on the Atlantic coast, which has the black band on the neck completely encircling it, and this has been named *A. m. circumcinctus*, but it differs in no other way.) Bill yellow, black at tip. Legs orange-yellow. A colored ring around the eye. Length about 6½ inches. Extent about 14 inches, being a trifle smaller than No. 102.

They are very pretty birds to shoot, but, if wounded, will run like a witch, and are then a good test of marksmanship.

104.   SNOWY PLOVER.

Ægialitis nivosa.

A little lighter in color than No. 103, and with a reddish tinge on the top of the head, which is nearly surrounded by a band of black. There is also a patch of black on either side of the neck, but it makes no attempt to encircle it as in No. 103, and there is no black band on breast as in No. 105. Bill black and slender. Legs black. Length 7 inches. Extent about 14 inches. Eyes and eyelids dark brown.

This bird is found upon the Western shores and sometimes about the Gulf of Mexico and the salt lands of the interior. I know nothing of its habits, but am informed that it resembles No. 102.

105. WILSON'S PLOVER.

Ægialitis wilsonia.

This bird resembles No. 104 very closely, but differs in having a black band on the neck which does not extend over upon the back. The bill is very stout and large, and the legs are flesh-colored. Slightly larger than No. 104; that is, a small specimen of this species would be about the same size as a large specimen of the other.

It is found quite commonly on the shores of the South Atlantic and Gulf States.

106. MOUNTAIN PLOVER.

Ægialitis montana.

Prevailing color grayish-brown. Entirely white beneath. Forehead white, bordered with black. In winter, or in the young birds, the black marks on the head are not present, and the plumage has a rusty tinge. Bill black and slender. Legs pale brown. Length about 10 inches. Extent 18 inches. Eyes dark brown.

These birds inhabit the high prairies, and seldom are found near marshy lands. They feed upon insects.

107. TURNSTONE. CHICKEN PLOVER. CALICO-BACK.

Arenaria interpres.

The adult male in breeding plumage is dressed like a harlequin, the back being blotched with black, white, brown and chestnut. The under side is pure white, broken by a broad patch of black upon the breast. The lower part of the back is white, with a black patch on the rump, and the tail is black bordered with white. The wings also, bear a large white patch. The head is mottled black, brown and white with a broad black patch on either side extending down to the sides of the neck and connecting with the breast patch. The colors of the female are similar but the chestnut color is replaced with brown, and the general effect not so bright. Immature and winter plumage mostly brown and gray. The bright colors of the back are generally a distinguishing feature sufficient for recognition. The bill is black. Feet orange. Eyes black. Length 9 inches. Extent 18 inches. There is a variety on the Pacific coast in which the characteristic red color is replaced by black.

These birds are found in nearly all quarters of the globe, and generally in small flocks of three to twelve, feeding upon the marine animals which it collects upon the shores by turning over the small stones, whence its name. Their favorite haunts seem to be

the pebbly beaches, but they are often taken
upon the marshes and upon the low sand
flats and bars.

They are one of the most conspicuous shore
birds, and once seen will be easily recognized.
Their flight is very rapid, and, while on the
wing they incessantly repeat their short,
sharp, whistling note, which is especially
hard to imitate, but they are not very shy,
and will come down to almost any of the
shore-bird decoys. I never tried to eat them
but once, and was not favorably struck by
their flavor, which resembles that of fish oil.

108. AMERICAN OYSTER CATCHER.

Hæmatopus palliatus.

Predominant color smoky-brown, with
black head and neck. Under parts white,
and the wing bears a conspicuous white patch.
(There is a variety on the Pacific coast, *H.
niger*, which is entirely black.) Bill 3 to 4
inches long, varying in every specimen, and
often thin on the end and bent to one side,
from its constant use in opening the shells of
mollusks; coral red, yellow tipped. Feet
and legs livid. Eyes and eyelids red.
Length 18 to 21 inches. Extent 30 to 36
inches. Said to be non-edible, although I
can see no reason why it should be so, as
they feed almost entirely on shell-fish.

E. H. Forbush, State Ornithologist of Massachusetts, tells me that this bird is very good eating. if quickly cleaned. And it is probable that many of those birds, which are ordinarily not esteemed fit for food, if drawn as soon as killed, and soaked a little while in water, would be found very palatable.

LAND BIRDS.

109. BOBWHITE. QUAIL.
VIRGINIA PARTRIDGE.

Colinus virginianus.

Predominating color reddish-brown, mottled and streaked with black and darker brown, lighter on the under parts. The male has the forehead, line over the eye, and throat white, bordered with black, which color is replaced in the female by buff. Bill dull black. Legs gray. Eyes dark brown. Length 9½ to 10½ inches. Extent 14½ to 15½ inches. Weight about 7 ozs.

There are two so-called varieties of this bird. Its range extends from Massachusetts to Texas and westward throughout the eastern United States. The bird described above is the one from the northern section. As we follow the species southward it gradually becomes smaller and darker colored on the eastern coast, producing the Florida variety (*Colinus virginianus floridanus*), and in the extreme southwest, becomes more gray, which variety is called *C. v. texanus*, but they are all practically bobwhite just the same.

There is no game bird more universally known and admired, for to the sportsman he affords a tempting mark as he flushes before the dogs, and whirrs off through the sharp air of autumn, his little body blending closely with the gray-brown of the fading foliage, to drop, after a short flight, in the first convenient covert, where he is soon located by the keen scent of the dog, to fall perhaps, at the gun's report. The bobwhite frequents the semi-open fields and pastures which afford him cover in the short brush and food in the open. At night, clustered together, the flock passes the sleeping hours, and during bad weather, in these same covers, and often in severe storms of sleet, they are frozen under by the accumulation of the frozen snow, to miserably perish. I know of none of our game birds whose very existence is so constantly in danger; a prey to the prowling fox by night, and chased and harassed by birds of prey by day, added to the other evils of destruction. It is a wise provision of Providence that they are so prolific, a single hen bringing up each year two litters of ten to fourteen chicks. They feed upon insects and the seeds of weeds, and no better ally can the farmer have for the protection of his crops than a few coveys of quail in his fields.

110. MOUNTAIN PARTRIDGE.
PLUMED QUAIL. MOUNTAIN QUAIL.
Oreortyx pictus.

I shall not here attempt to separate the two
varieties now laid down in the list, as they
are practically the same bird. Predominating
color olive-brown, with a coppery lustre.

Breast slaty-blue, shading into the olive at the
back, and finely marked with black. Throat
and belly chestnut. There is a black line,
bordered above with white, on either side the
neck, and the sides are banded with broad
bars of white and black. The distinguishing
feature of this bird is the two black arrowy
plumes on the crown of the head, which are
3 to 4 inches long in the male and much
shorter in the female. Bill and feet brown.
Length 12 inches. Extent 17 inches.

This beautiful bird is an inhabitant of the
mountainous regions of the northwest, haunt-
ing the underbrush. They are not fast flyers,
and will often try to skulk off when approached, which tactics are extremely suited

to the thick covers which they frequent. The opinions of my various correspondents differ as whether they will lie well to the dog, some say yes and some no, but all agree that they furnish fine sport.

111. SCALED PARTRIDGE.
BLUE QUAIL.
Callipepla squamata.

Predominating color leaden-blue, darker on the back. The distinctive feature of this bird is the peculiar shell-like marking of the neck and breast, produced by the black edgings of the feathers. The crest is short, and composed of several feathers, but is not so prominent as in others of the crested quails. There is a large patch of reddish-brown on the belly, which in some specimens shades into chestnut, giving rise to a variety known as *castanogastris.* This latter variety inhabits the lower lands; and the main species, the tablelands of Texas, New Mexico, and Arizona. Length 11 inches. Extent 15 inches, being a little smaller than No. 110.

112. CALIFORNIA PARTRIDGE.
VALLEY QUAIL. (var.)
Callipepla californica.

Predominating color ashy, shaded with olive-brown. The breast is bluish-slate; below this, on the under side, is a patch of chestnut, shading into tawny brown toward the edges, with the tips of the feathers edged with black. The head of the male is marked with a white line over the eye and a white forehead. The head is surmounted with a crest of 6 to 10 black feathers, curved forward.

The female has not the black throat nor the head markings, and the crest is smaller.

Length 11 inches. Extent about 16 inches. It inhabits California and Oregon east to the Colorado River.

Much has been written concerning the comparative merits, as a game bird, of this, as compared with our eastern quail, and, from what I can gather, I am inclined to think that bobwhite is on top. The Californian does not lie well to the dog until they have been badly scared, and then they go like a "blue streak." However, they are a good table bird, and afford considerable sport, since they are quite numerous. They habit the thick cover of chaparral and weeds.

113. GAMBEL'S PARTRIDGE.

Callipepla gambeli.

This bird is own cousin to the last (No. 112). differing in having a black forehead instead of white; no white line beneath the eye; back of head chestnut instead of smoky-brown; sides chestnut with white stripes; and the middle of the belly jet-black instead of chestnut. Otherwise in appearance and habits like its Californian relative.

114. DUSKY GROUSE. BLUE GROUSE.

Dendragapus obscurus.

Predominating color very dark brown, finely marbled with gray, shading into bluish-gray on the under sides. Cheeks black. Tail rounded, of 18 to 20 feathers, dark brown like back, tipped with a band of gray. Bill black. Eyes golden-brown, with a comb above. Length 20 to 24 inches. Extent about 30 inches. Weight 3 to 4 lbs.

The female is of a lighter shade than the above, and a little smaller than the male.

This great bird is a native of the Rocky Mountains, the darker varieties known as *Richardson's* and *Sooty Grouse* occupying the northern limits.

From all that I can gather from my correspondents in relation to its habits, it seems to be a somewhat stupid, lazy bird, not easily flushed, but a strong, rapid flyer, when startled; frequenting the high coniferous trees during the most of the year. From all accounts, it affords but comparatively poor sport, aside from its great size and abundance, as it must be shot while sitting, and it appears to have the ability to so adjust itself upon the limbs as to almost completely hide from the gaze of the observer. One correspondent tells of killing six by throwing stones at them, and another of shooting them out of the high trees 100 feet from the ground. In some places they are shot when flushed, and, in the other instances, they should afford good sport with a rifle.

115. CANADA GROUSE.
SPRUCE PARTRIDGE.

Dendragapus canadensis.

Prevailing color of adult male, black, lighter on the back, waved and spotted with white and tawny. Tail of 14 to 18 feathers tipped with brown. The female and immature male are somewhat like the above on the back, but not so dark, and the under side is variegated white and tawny. waved with blackish. There is a red comb over the eye. Length about 16 inches. Extent 22 inches. There is a slightly differing variety in the Northwest. The habitat of this bird is the northern part of the continent, south into the northern border of the United States.

Although a very handsome bird, it is little sought for by sportsmen, since its flesh, though sometimes not bad, is generally so impregnated with the flavor of the spruce buds upon which it feeds, as to render it utterly uneatable. And it affords but little sport, for it is so stupid as to allow the approach of man within a few yards, and is even taken with a slip-noose on the end of a pole.

I saw three of these birds breeding in captivity at Kentville, N. S., and I believe that their owner, Mr. Bishop, enjoys the unique honor of being the only person who has succeeded in their domestication.

116. RUFFED GROUSE.
PARTRIDGE. PHEASANT.

Bonasa umbellus.

Prevailing color variegated grayish-brown (in some specimens shading into a reddish-brown with bronze lustre). Whitish below, barred with brown.

The male has a ruffle of glossy black feathers about the sides of the neck, which in

the female is smaller and of a brown color.
The tail, normally of 18 feathers, is rounded,
and bears a band of black near the tip. There
are three varieties, due to climatic agencies,
and they grade into each other insensibly.

Bill and legs light brown. Eyes brown.
Length 16 to 20 inches. Extent 23 to 25
inches. Weight 1¼ to 2 lbs. I have heard
of "King-partridges" as large as a turkey,
but never saw one: and a 2-pound bird is a
big one.

The birds inhabit the greater portion of
civilized North America, except the Southwest, and, except where seldom hunted, are,
taken all in all, the most noble game bird of
which this country boasts. I have seen these
birds flush from the ground and settle upon
a tree not over 200 feet away, and eye me as
curiously as if I was a dime museum freak,
and I have been threatened with condign
punishment by correspondents in the North
and West for the above remark. But if any
of these friends will have a try at these birds
here in our Eastern States, I will wager that
they change their minds. The sportsman
who brings in a good bag of these birds here
is worthy of a place in the front ranks of the
guild. It is warier, and filled fuller of strat-
egy, pound for pound, than any living bird on
our soil.

Like nearly all gallinaceous birds, the male
has the habit of strutting during the sexual
period, which is accompanied by various

sounds calculated to attract the female and exalt it in her estimation; and, in the case of the ruffed grouse, the act is termed "drumming." At this time they seem to be oblivious of all else, but he is a good man who can surprise the bird even then.

Any discussion of the habits of the bird would seem superfluous; volumes have been written on the subject, and more will follow just as long as man can walk the fields.

117. WILLOW PTARMIGAN.

Lagopus lagopus.

There is no American bird which exhibits such peculiarities of plumage as do the Ptarmigans. For they are constantly moulting, and there are no two weeks in the year when a bird would correspond to any accurate description; and no bird exhibits such a complete change, for even the horny coverings of the bill and toes are cast off with the feathers, and the plumage varies in every gradation of color, from the beautifully mottled summer coat of blacks and browns to the nearly uniform white of winter.

In the full spring breeding plumage, the predominant color is rich brown, inclined to tawny, mottled and barred with black and white; the most of the wings and the abdomen white. The female is similar, but more tawny, including the abdomen. The legs are feathered to the toes; the winter plumage is

white, except that the tail is black. Ornithologists have lately divided this species into
two. The principal point of difference being
that the original species has the quills of the
outer wing feathers white, and the variety
(*alleni*) black. Bill black and stout. Eyes
hazel. Length about **17** inches. Extent
about **24** inches.

This bird takes the place in the North of
the grouse in the South, and they afford as
fine sport in their haunts as do their relatives
of the South. They are strong flyers, and lie
well to the dog.

They are found on the marshy lands, and
their numbers, in those sections where they
have not been too much molested, is something incredible, their cry of "Go back, go
back" resounding from every quarter when
disturbed.

118. ROCK PTARMIGAN.

Lagopus rupestris.

This bird much resembles the foregoing,
but is smaller, and the color is more of a
brownish-yellow, and in the winter there is a
black stripe on each side of the head, which
is not possessed by No. **117**. The bill is more
slender. Length **14** to **15** inches. Extent
about **22** inches.

There are three varieties of this species
described by the later ornithologists, and another bird from Newfoundland (*L. welchi*)

which has been considered sufficiently distinct to form a new species. But they are sufficiently similar to be considered as one and the same, so far as our purpose goes, and in view of the numerous variations of individual plumage, the subdivision is extremely problematical, except to the expert.

119. WHITE-TAILED PTARMIGAN.

Lagopus leucurus.

Habitat — Rocky Mountains. Size and appearance similar to No. 118; but the tail remains *white* throughout the year, and the wings and abdomen are of this color in the summer plumage. Found on the mountain ranges, from the timber-line upward, coming lower down in the winter, as the other forms come southward at the same time.

120. PRAIRIE HEN.
PINNATED GROUSE.

Tympanuchus americanus.

Predominant color grayish-brown, heavily barred with black. Head buffy, underside dirty white, barred with brown. Top of head and stripe on each side, black. Tail dusky, tipped with white. The most peculiar feature of this bird is the black tuft of feathers which adorns the neck on either side, beneath which is a large, bare patch of yellow skin which in the breeding season the male distends. The female is very similar, but the neck tufts are shorter. Length 17 to 18 inches. Extent 28 to 29 inches and over. Eyes brown. Bill brown. Feet yellowish.

This bird, while not for a moment to be classed with the ruffed grouse, is still a great favorite with sportsmen. Their flight is not so rapid as that bird, but more protracted and unaccompanied by the disconcerting whir-r-r of the other, although they can raise up quite a racket when suddenly disturbed.

Their flesh is dark, and does not compare favorably with that of the ruffed grouse, but being more plentiful they occupy a favorite place in the market.

Their habitat is the open prairie, seldom visiting the timber. In former days, this bird, or a representative of the family (the heath-hen), was not uncommon on our Eastern plains, but they have gradually dwindled away, until only one colony now exists, upon the island of Martha's Vineyard. And such is the greed for gore and gain, that were it not for the strenuous efforts of a few far-sighted men, who can see more than a foot beyond their noses, these birds, too, would have long since gone the way of the Great Auk and the Labrador Duck. I cannot speak in too strong words of contempt of the action of those, who, even now in spite of the protecting laws, seek to remove the last of this race which once was found commonly distributed over our land.

There is a variety of the prairie hen which is found in the South, smaller and lighter in color. The characteristic bars are narrower, and brown or grayish, rather than black. The length is 15 to 17 inches, and the extent 27 inches. Otherwise it differs, neither in habits nor appearance, from the usual form as commonly seen. In fact, it is merely a form, which, by reason of its surroundings, has become changed from the original type, and being isolated from it, has in progress of time become perpetuated.

121. SHARP-TAILED GROUSE.

Pediocætes phasianellus.

This is the northwestern representative of the prairie-chicken, and is somewhat similar in habits, but not at all in appearance. The predominant color is ashy-gray, marked with irregular spots and bars of black, white and tawny-brown. The underside is nearly white, and marked with U-shaped spots of brown. The throat is light buff color, and there is quite a pronounced crest on the top of the head. Unlike the preceding species of grouse, this bird has no neck ornaments, but a close examination will disclose there the characteristic bare spots, which are seen during the breeding season. Two varieties have been made of this species. The Northern form is darker, with but few brown markings, and the spots on the breast are V-shaped; and, curiously enough, the lighter and browner form, of the extreme South, also exhibits this V-shaped mark, although separated from the other by the common type. A distinctive feature of this bird is the tail, which is triangular in shape, when in a natural position, the middle tail feathers being about four times as long as the outer. From this arises the name. Bill and feet brown. Eyes brown. The legs are feathered to the feet. Length 18 to 20 inches. Extent 25 to 30 inches.

As compared to No. 120, this bird ranks high as a game bird. Its flight is strong, swift and direct, and its flesh much superior to the other bird.

An attempt, undoubtedly so far successful, is being made to introduce this bird to our eastern fields. While it will probably never take the place of our ruffed grouse in the estimation of our sportsmen, any addition to our feathered game will be welcomed, to afford a bag to those who desire to hunt over our fields, now rapidly being depleted of game birds.

122. SAGE GROUSE. SAGE COCK.
Centrocercus urophasianus.

The monarch, so far as size is concerned, of the grouse, but, alas! there his dominion ceases. Many times the tyro sportsman has been sadly left, when he has thought himself possessed of a good dinner in an old sage cock. A correspondent writes: " I never attempted but once to eat an old bird, and I shall never try again, I would rather eat a broiled plug of tobacco ; but for a fine morsel, commend me to a young bird in the summer. I could not believe such a transition possible; they beat a woodcock 'all hollow ' at that time." The color is gray, variegated with black, brown and buff on the back, and dirty white below, and the neck has two bunches of hair-like feathers, beneath which is the air sac which can be extended to enormous proportions. The tail is long and composed of twenty stiff and narrow feathers. The full-grown cock is over 2 feet long, while the hen will seldom reach this measure. The extent is about 3 feet. with about the same proportion for the female bird. Weight 3 to 5 lbs.

When this great bird gets up in front of the gunner, he is apt, if not accustomed to the bird, to have a severe attack of ague, although one friend tells about kicking them out of the sage-brush. The flight is strong, and when a wild bird is started, don't try to follow him, he may go a mile.

123. WILD TURKEY.

Meleagris gallopavo.

There are two species of this bird in North America, else I should deem it unnecessary to give any description. The wild turkey proper was formerly distributed generally over Eastern North America, but it is now found only in those portions but little visited by man, and not even there are they, by any means, common. The general color of this bird is black, with a coppery lustre, each feather being margined with velvety black. The tail feathers are dark chestnut, with numerous bars of black. This same color characterizes the sides and coverts. The southern species, which is the original of the domesticated turkey, has the feathers at base of tail, on the back, chestnut, tipped with whitish, and the tips of the tail feathers are whitish. In both species the head is bare of feathers, colored blue, with reddish excrescences. The forehead is furnished with a depending, fleshy, cone-shaped process, which is erected in moments of excitement. There is also a hairy tuft upon the breast. Length 3 to 4 feet. Extent 4½ to 5 feet and over. Eyes brown. Bill brown. Legs dark and reddish-purple.

This monarch of game birds, like similar representatives of human bipeds, is rapidly disappearing before the march of civilization and progress. Once common in New England, it is there extinct; and this noble bird must now be followed by sportsmen to the forest fastnesses of the South and Southwest. A few still exist in the Allegheny range, but they are seldom taken.

A successful turkey hunter is the embodiment of all that appertains to woodcraft; and he who can call a gobbler within range is worthy of a place in the front ranks of the sportsman fraternity.

Their food is a mixture of nuts, seeds, and insects, and their flesh is as much esteemed for the table as that of the domestic bird.

124. PASSENGER PIGEON.

Ectopistes migratorius.

Color blue above, reddish-brown beneath, becoming lighter toward the tail, which, composed of 12 feathers, is brown in the middle and blue on the sides. These blue feathers, when pulled apart, show a web of white on the inner sides. The neck is beautifully glossed with a golden-violet. Bill black. Legs bluish. Feet red. Eyes red. Length 16 to 17 inches. Extent about 2 feet.

This is the bird popularly known as the wild pigeon. It was in former years commonly distributed over the entire country, passing North and South during their migrations, in immense numbers. Even now, scattering birds are seen in the East, but we must now go to the West to find the flocks, and, there even, their numbers are greatly lessened.

It could hardly be called sport to hunt them when they were common, for a shot fired into the midst of a flock would bring down numbers of them, but they are now somewhat followed as game. They are fine eating; and it takes a good shot to secure a single bird, as their powers of flight and its velocity are wonderful. They migrate in flocks, sometimes of immense size, but although they remain in colonies, they breed in single pairs.

INDEX.

www.ingramcontent.com/pod-product-compliance
Lightning Source LLC
Chambersburg PA
CBHW021823190326
41518CB00007B/718